Population Neuroscience

Tomáš Paus

Population Neuroscience

Springer

Tomáš Paus
Baycrest
Rotman Research Institute
Toronto, ON
Canada

ISBN 978-3-642-43776-2 ISBN 978-3-642-36450-1 (eBook)
DOI 10.1007/978-3-642-36450-1
Springer Heidelberg New York Dordrecht London

Printed on acid-free paper

Springer is part of Springer Science+Business Media (www.springer.com)

Preface

For centuries, we have wanted to know how the human brain works. How do we build, store, and retrieve memories? How do we understand spoken and written language? Are such different functions localized in distinct brain regions that differ in some special properties? If so, what are those properties?

In the past 50-plus years, we have been able to answer many of these questions and, in general, have gained an amazing amount of knowledge about the structural and functional organization of the human brain. In the process, we have also started asking questions about distal causes of the apparent variability manifested by *healthy* brains. Do the brain cells of a pianist differ from those of a person who has never played a musical instrument? Is Einstein's brain different from Bach's? And what about the mother's diet during pregnancy: does it support the growth of certain elements in the brain without affecting others? And peers during adolescence: do they leave a long-lasting signature in the social brain? Does the way we live in our middle years affect how our brains age?

To answer these and many other questions, we can now turn to population neuroscience. With in vivo imaging we are able to ask what shapes the brain, both from within (genes) and from without (social and physical environment). There are two main reasons for doing so. First, it is likely that by uncovering sources of inter-individual variability in the (healthy) human brain, we will acquire knowledge about *processes* leading to a particular *state* of brain structure and function. Second, by gaining insights into a process, we get closer to *prediction*. This is because individuals with a particular constellation of distal causes, and the ensuing developmental cascades, will likely differ in the state of their brains at a particular time in their life and, by extension, in their risk for developing a brain disorder. Thus, over the long term, understanding distal causes and associated processes will lay down the foundations for personalized preventive medicine.

Can these goals be achieved? The human brain is a large and complex organ: 1,300 cm^3 of tissue, 100 billion neurons, and 176,000 km of axons. Our genome is equally complex: ~20,000 genes with tissue-specific patterns of expression, regulated through multiple mechanisms, including epigenetic ones. Add to these a mind-boggling number of possible social and physical environments the individual encounters from conception onwards—further affected, perhaps, through environment-induced epigenetic modifications of his/her ancestors' genes—and clearly we have a very complex array of factors in the shaping of a human brain.

How can we deal with such a level of complexity? As outlined in this book, one possible approach is that of population neuroscience, where detailed information about the individual's envirome, genome and epigenome, and (brain) phenome is collected simultaneously in a large, population-based sample. Subsequently, these multi-level datasets are combined in analytical models to gain new knowledge about the process, thus enabling prediction of the individual's risk of a brain disorder. By its nature, this work can be carried out only by teams of scientists with a wide range of expertise, including epidemiology, genetics, cognitive neuroscience and brain imaging, engineering, computational science, bioinformatics and mathematics, as well psychiatry and neurology, behavioral sciences, law and ethics. By introducing the relevant vocabulary, concepts, and tools, this book aspires to prepare students and practitioners of these diverse disciplines to join such multi-disciplinary teams.

The book is laid out as follows. In Chap. 1, the reader will learn about the key terms and concepts used in epidemiology (e.g., "exposures" and "outcomes"), genetics (e.g., genome, transcriptome, proteome), and cognitive neuroscience (regional specialization, structural and functional connectivity, development and plasticity). We will also introduce the concepts of envirome and phenome, and point out the needs for triangulation in enviromics and multi-level high-throughput approaches in phenomics. The reader will then learn about developmental cascades as a model for thinking about multi-factorial traits emerging over time. This chapter will end by introducing the concept of personalized preventive medicine.

Next, in order to help practitioners of one discipline understand the ways of thinking common in another, Chap. 2 will provide brief historical accounts of epidemiology, genetics, and cognitive neuroscience, the three disciplines that form the foundation of population neuroscience. In the five chapters that follow, the reader will learn about diverse measurements and tools used to acquire them, which are available to researchers in order to characterize an individual's envirome (Chap. 3), genome(Chap. 4), epigenome (Chap. 5), molecular phenome (Chap. 6), and systems-level (brain) phenome (Chap. 7).

Chapter 8 will describe in brief a number of population-based imaging studies, in order to illustrate the use of some of the tools mentioned in the previous chapters; we will consider here both birth cohorts and prospective population-based studies initiated during childhood and adulthood. In Chap. 9, the reader will discover some key challenges and their possible solutions. We will review ways in which we can construct hypotheses through meta-analyses, address the key issue of association versus causality and discuss the need for going beyond the "MR brain".

Finally, Chap. 10 will return to personalized preventive medicine and speculate about possible ways of using the knowledge gained in population-based studies of the human brain for moderating the individual's risk and resilience vis-à-vis common brain disorders. We will talk here about healthy life expectancy, the cost of common chronic disorders, risk profiling, and possible tools for transmitting personalized high-tech information to the "client." We will close with a hypothetical scenario that points out the social and economic costs of Alzheimer's Disease and offers possible solutions for delaying the onset of this disease through personalized preventive medicine.

This book brings together knowledge gained through many years of my work on the human brain, first as a cognitive neuroscientist and later as a population neuroscientist. Over the past 30 years, I have been fortunate to have wonderful mentors and colleagues—starting with Miloš Kukleta during my medical studies in the Czech Republic, going on to Brenda Milner, Michael Petrides, and Alan Evans during my years at the Montreal Neurological Institute, and also many colleagues on both sides of the Atlantic, who worked with us on a number of MR-based population-based studies of the brain and body. My students and fellows have inspired me to find answers to their questions—thus pushing me to a clearer understanding of my own words and concepts. Their input has been invaluable and I am indebted to them for it.

This book would not have been possible without Zdenka Pausova, my partner in both work and life. Over more than 20 years, Zdenka has provided me the inspiration and knowledge necessary for embarking on studies in genetics and epidemiology. Together, we built the Saguenay Youth Study, which provides the template of most of the ideas and concepts described in this book. And I also want to thank our daughter, Veronika, for her positive energy and patience with the "science talk" while driving to school or camping.

Finally, I would like to acknowledge my gratitude to the Tanenbaum family for supporting population neuroscience at the University of Toronto. Thank you for investing in our work and sending the message that this research can have a major impact on the future health of our children and grandchildren. And, last but not least, I am thankful to Rosanne Aleong, Amanda Celis, Anja Klasnja, Mamta Pranjivan, and Deborah Schwartz for their assistance with figures, tables, and references, and to Louise Fabiani, a science writer form Montreal, for her careful reading of the drafts of this book and her suggestions for improving it.

Toronto, November 30, 2012 Tomáš Paus

Contents

Chapter 1
Terms and Concepts

1.1 Epidemiology: Exposures and Outcomes

Every phenomenon has both distal and proximal causes. Take the example of an accelerating car. A distal cause would be the driver pressing harder on the gas pedal. A proximal cause could be the increasing frequency with which the engine's cylinders move up and down or the increasing speed of the car's wheels.

In biology, we do not often find such linear chains of causes and consequences. Nonetheless, it is helpful to assign roles to the various measures when investigating sources of the inter-individual variability in a particular outcome of interest. Typically, we may refer to factors originating in an individual's external environment as exposures and their consequences as outcomes. In a sense, this exposure–outcome dichotomy invoked in epidemiology is similar to that between independent and dependent variables used in psychological experiments, where we manipulate the values of an independent variable (e.g. sharpness of a visual stimulus) while measuring those of dependent variables (e.g. speed of the individual's response to the stimulus). In real life, however, we *observe* rather than manipulate exposures, hence the term observational epidemiology. Furthermore, many exposures are not truly independent of the individual's behaviour; for example, the degree of a person's exposure to infectious agents during a flu epidemic would depend on his/her avoidance of public spaces. The fact that we do not manipulate exposures in observational studies poses a challenge with regards to interpretation of the observed *associations* between exposures and outcomes. As a result, we are unable to attribute causality to the exposures.[1]

Instead, we may explore possible roles of the measured variables by setting up theoretical models and testing them through statistical means. Using such models, we are trying to explain (or predict) inter-individual variability in one variable by another variable; we do so in various pairwise combinations while taking into account variations in the third (fourth, etc.) variable. In this way, we are attempting to characterize "transactions" between different variables in order to assign one of

[1] This issue is discussed in detail in (Sect. 9.2).

T. Paus, *Population Neuroscience*, DOI: 10.1007/978-3-642-36450-1_1,

the following roles to each of them: exposure, outcome, mediator, moderator and confounder.

A good example can be found in the possible relationship between breastfeeding and general intelligence. As a thought experiment, let us consider a possible prospective study in which we measure the duration of breastfeeding, head circumference and general intelligence at 7 years of age, as well as the volume of the brain's white matter and the participants' general intelligence at 17 years of age. We also know about a gene involved in the synthesis of fatty acids from breast milk and have information about which of the three variants of this gene (genotype) each of the participants carries. Finally, we wish to rule out the (confounding) influence of a number of factors likely associated with general intelligence (and/or breastfeeding), such as maternal education.

As shown in Fig. 1.1, the use of statistical modelling for our hypothetical data allowed us to assign the following roles to the variables listed above. By design, breastfeeding is the *exposure* and general intelligence is the *outcome*. Maternal education is a *confounder*. Head circumference at 3 years of age predicts general intelligence at both 7 and 17 years of age and, therefore, can be considered a *mediator*; in this context, head circumference likely acts as a proxy of brain size, but—to strengthen this inference—one would need to consider as another

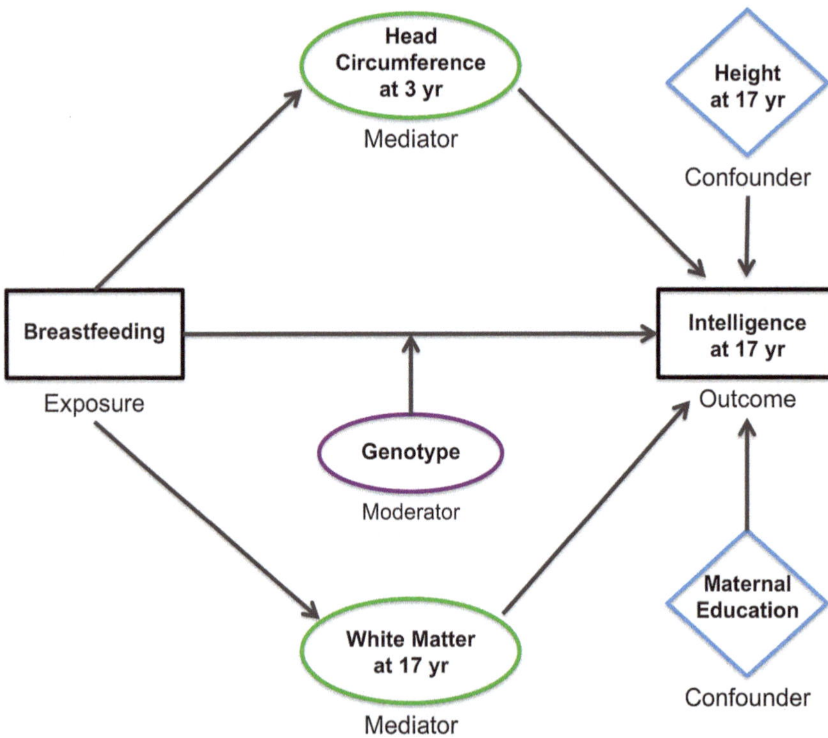

Fig. 1.1 Breastfeeding and general intelligence: mediation and moderation

confounder the overall growth of the body (e.g. height measured at 7 and 17 years of age). Importantly, the volume of white matter at 17 years of age is an excellent predictor of general intelligence at the same age; hence, we assign the *mediator* role to this variable as well. But the latter relationship holds true only for individuals with a particular genotype; thus, the genotype is a *moderator* of the relationship between breastfeeding and white matter. Finally, we also see a similar moderating role of the genotype vis-à-vis general intelligence at both 7 and 17 years of age, but no moderating effect on the relationship between breastfeeding and head circumference.

From this pattern of "transactions", we conclude that *exposure* (breastfeeding) affects the *outcome* (general intelligence) by influencing *mediator 1* (brain growth inferred from head circumference) and *mediator 2* (white matter), with one *moderator* (genotype) influencing the impact of exposure on mediator 2. This conclusion is reached after ruling out the possible role of two *confounders* (maternal education, height). Importantly, the moderating effect of the genotype hints at a possible mechanism underlying the relationship between breastfeeding and general intelligence: the role of fatty acids in the development of white matter.

In summary, observational epidemiology represents the first step in our quest to explain sources of inter-individual variability: it reveals novel associations between external exposures and outcomes and, through statistical means, sorts out possible mediating, moderating and confounding influences of other variables on the outcome. If we add information about the individual's genes, the shift from classical observational epidemiology to molecular epidemiology begins (see Sect. 2.1).

1.2 Genetics: Genome, Epigenome, Transcriptome and Proteome

The nucleus of each cell in the human body contains 23 pairs[2] of chromosomes made up of bundles of deoxyribonucleic acid (DNA) and proteins (histones), which allow DNA to be packaged into chromatin (a combination of DNA and histones). The human genome is coded by the sequence of nucleotides (adenosine, guanine, cytosine and thymine): A, G, C and T comprise the alphabet of the genetic code.

Genes (~20,000 of them) are the units of inheritance. Each gene consists of two alleles, of one DNA strand each: one inherited from the mother and another from the father. Genes are "expressed" when DNA is transcribed into a messenger RNA (mRNA), which leaves the nucleus and is translated into proteins on ribosomes (Fig. 1.2). Gene expression is regulated by a multitude of factors. Perhaps surprisingly, external exposures (e.g. diet, stress) may also affect gene expression via so-called epigenetic modifications. Such epigenetic modifications will often "switch off" a gene for the rest of a person's life.

[2] This is true about somatic cells, only. Gametes (sperms and eggs) contain only one set.

The genome is the map of genetic variations across all chromosomes. In a given individual, these variations—for example, in a single-nucleotide polymorphism (A or G)—are the same in all of his/her cells (and tissues). On the other hand, the epigenome is the map of epigenetic variations (methylations, acetylations) found in the DNA of cells in *specific* tissues. Thus, the epigenome of brain cells will not be the same as that of liver cells. This is important for population-based studies: while genetic variations observed in DNA extracted from blood cells (the most readily available biospecimen in human studies) are the same for all tissues, epigenetic variations observed in blood cells may or may not be the same as those we could find in, say, brain cells (which we cannot access in our living participants, for obvious reasons).

The same is true about the transcriptome and proteome; each cell type (and tissue) will have its own "map" of gene expression (transcriptome) and proteins (proteome) synthesized in that tissue. Furthermore, transcriptomes and proteomes are *dynamic*: when quantified, the mRNA and protein levels of a given gene will depend on a variety of factors, such as the time of day or stress level at the time of tissue removal.

As you can see, we can conceptualize variables derived from a genome as "independent" variables (exposures or moderators), while those derived from a transcriptome and a proteome are best thought of as "dependent" variables (outcomes or mediators). Similarly, the methylation status of a gene can be used as

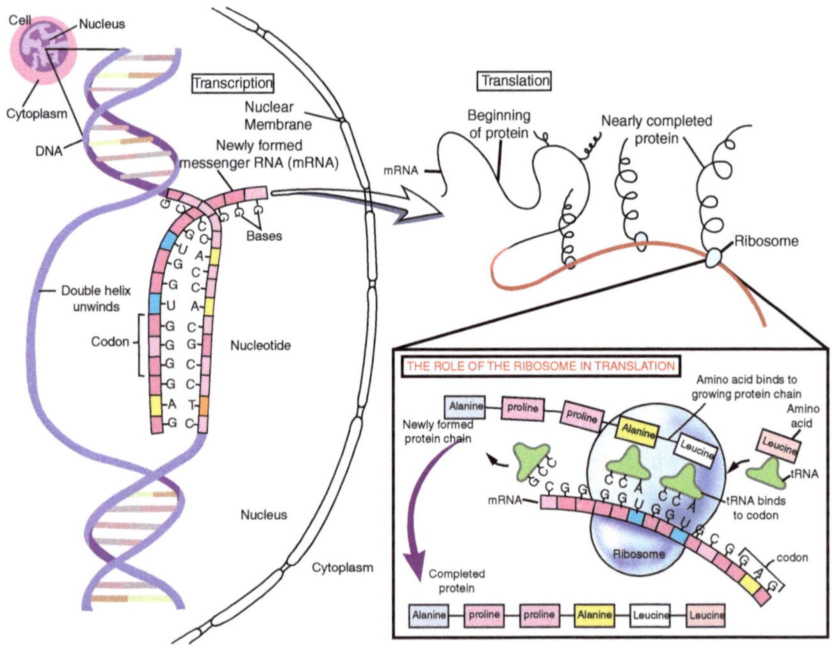

Fig. 1.2 From gene to protein. © 2001 Terese Winslow

an outcome of an external exposure (e.g. stress) or a mediator of the *effect* of stress on the expression of this gene.

Before we leave this section, let us touch on the origin of inter-individual variations in our genome and epigenome. In the case of genes, these variations emerge through two basic processes: (1) recombination during sexual reproduction and (2) *de novo* mutation in germ-line cells (oocytes and spermatocytes). Note that mutations occurring in the DNA of well-differentiated cells (in specific tissues, such as the brain or liver) cannot be passed onto the next generation. Similarly, epigenetic modifications of the DNA in germ-line cells can be passed onto the next generation, but those affecting specific tissue cannot. Thus, epigenetic inheritance is a vehicle for trans-generational transmission of environmental influences. In other words, the expression of a granddaughter's genes might in some case reflect her grandmother's lifetime exposure (Fig. 1.3).

Altogether, the "mappers" of the genome (genomics), epigenome (epigenomics), transcriptome (transcriptomics) and proteome (proteomics) provide us with a wealth of information about inter-individual variations in the molecules involved in *processes* that lead to a particular *state* of brain structure and function. Transcriptomics and proteomics, together with metabolomics and lipidomics, yield molecular phenotypes (see Chap. 6). Genomics reveals inter-individual variations in the pattern of nucleotides across our genome, such as single-nucleotide polymorphisms and copy number variations (see Chap. 4). And epigenomics opens up new avenues for exploring *de novo*, as well as heritable, environment-induced modifications of DNA, which affect gene expressions rather than the genetic code (see Chap. 5).

1.3 Cognitive Neuroscience: Specialization, Connectivity, Development and Plasticity

Population neuroscience is concerned with explaining the sources of inter-individual variations in the structural and functional properties of the human brain and,

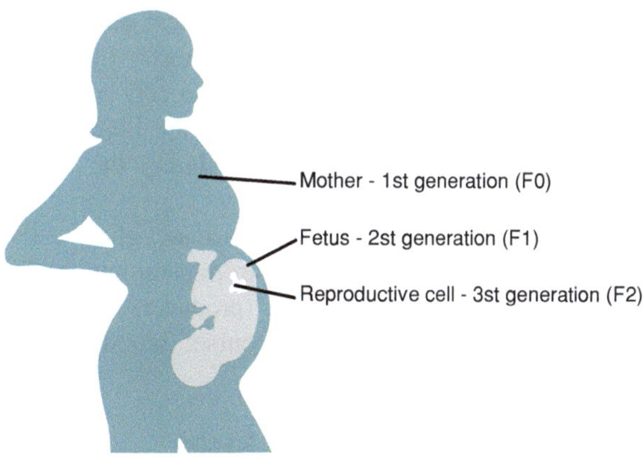

Fig. 1.3 The Russian doll of trans-generational epigenetics. Three generations at once are exposed to the some environmental conditions (e.g. diet, toxics, hormones). From Perera and Herbstman (2011)

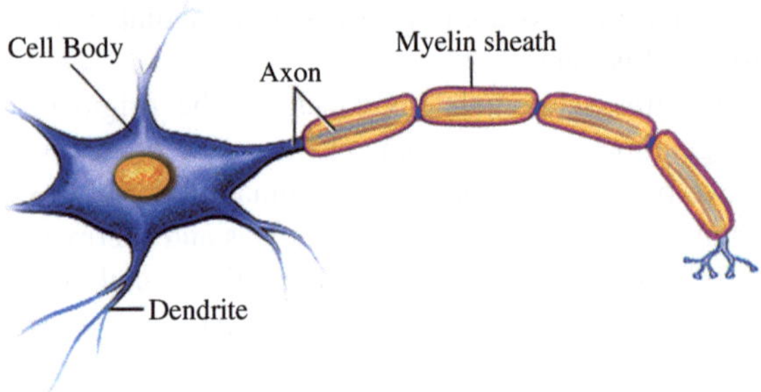

Fig. 1.4 Neuron

by extension, behaviour. In large studies, these properties can be most readily measured with structural and functional magnetic resonance imaging (MRI), together with cognitive and mental health assessments. These "system-level" phenotypes are described in detail in Chap. 7. Here, we will touch upon basic facts about the human brain, its organization, development and plasticity.

The human brain is an organ capable of a multitude of functions. It can see and hear, move and think, suffer and enjoy. It appears to function as a mosaic of highly specialized "modules" located in spatially distinct brain regions, which are interconnected to form neural networks. In these networks, information is computed in nodes (or modules) specialized in extracting particular features of the "reality", such as the speed, shape and colour of a moving object. To form a representation of the "reality", this information is shared across the nodes, new information is derived based on the combination of the individual features, an object is named ("apple") and finally an appropriate plan of action is formed and executed ("catch the apple").

The main phenotyping tool—brain mapping—operates in the three-dimensional space of this 1,300 cm^3 organ. It is therefore important to keep oriented in this maze of cells, tissues and locations.

Based on the appearance of the brain tissue to the naked eye, we can distinguish grey matter (~55 % of brain tissue) and white matter (~45 %). Grey matter consists of neuronal cell bodies and their short extensions (or processes) called dendrites. White matter is occupied mostly by the long processes called axons, which carry electrical signals from the cell body to the synapse (Fig. 1.4). The white matter owes its appearance to the high concentration of myelin, a fatty substance wrapped around and insulating the axons, thereby speeding up the conduction of electrical signals.

Interspersed between the neurons are glial[3] cells: astrocytes (important for metabolism and recycling of neurotransmitters), microglia (performing immune functions) and oligodendrocytes (producers of myelin in the brain). Also found between the neurons are, of course, blood vessels and neuropil—the latter

[3] The term "glia" comes from Greek for glue.

Fig. 1.5 The relative volumetric contribution of various cellular compartments in one cubic millimeter of mouse cerebral cortex. E, extracellular space. Data from Braitenberg (2001). From Paus (2009)

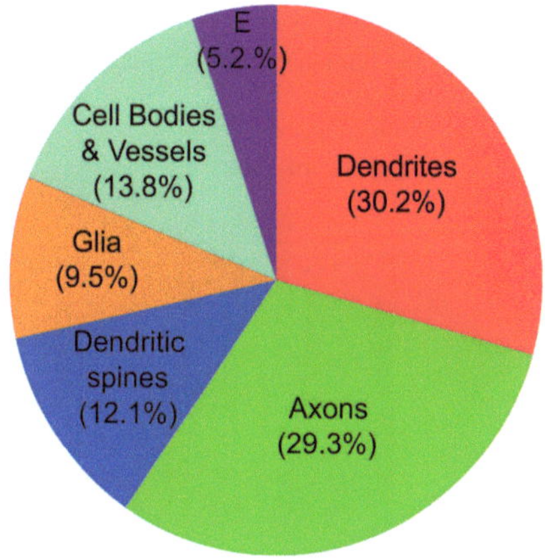

containing small neuronal and glial extensions (Peters et al. 1976)—as well as the fluid of extracellular space. Not surprisingly, one cubic millimetre of cortical tissue contains a mix of different cells (Fig. 1.5). We need to keep this in mind when interpreting brain phenotypes obtained with MRI.

The majority of grey matter is found in the cerebral cortex, a highly folded sheet of grey matter.[4] The rest of grey matter is buried in the depth of the brain, in structures such as the basal ganglia and thalamus, as well as in the "small brain" (cerebellum) located at the back (Fig. 1.6).

The grey matter of the cerebral cortex can be divided, somewhat arbitrarily, into four lobes: frontal, parietal, temporal and occipital. Anatomists have also named all the various folds (or sulci) of the cerebral cortex. With the naked eye, one can therefore locate a cortical region by its proximity to one of the sulci (e.g. the central sulcus) or one of the gyri, concave "bumps" next to the sulci (e.g. the precentral gyrus). Under the microscope, the cerebral cortex has been subdivided into a large number of areas based on their unique cytoarchitecture. In other words, neurons of different sizes and shapes are found in different densities in the six cortical layers. This cellular architecture appears to correlate with the local pattern of neurotransmitters and their receptors (chemoarchitecture) and, not surprisingly, with the local pattern of gene expression. Clearly, a distinct cellular phenotype (e.g. the size and shape of a cell and the neurotransmitters it synthesizes and responds to) is, to a great extent, the manifestation of genes expressed by a given cell.

White matter appears more homogeneous across the brain, both to the naked eye and under the microscope; only a few major bundles of axons (or tracts) stand out, such as the corpus callosum, which connects the left and right cerebral hemispheres (Fig. 1.6). New MR-based imaging techniques can segment out many other fibre tracts, based on the directionality of water diffusion. (we will describe these and other measures of structural and functional connectivity in Chap. 7). But just as we

[4] If unfolded, the ~3 mm-thick six-layered cerebral cortex would occupy an area of 2,500 cm^2.

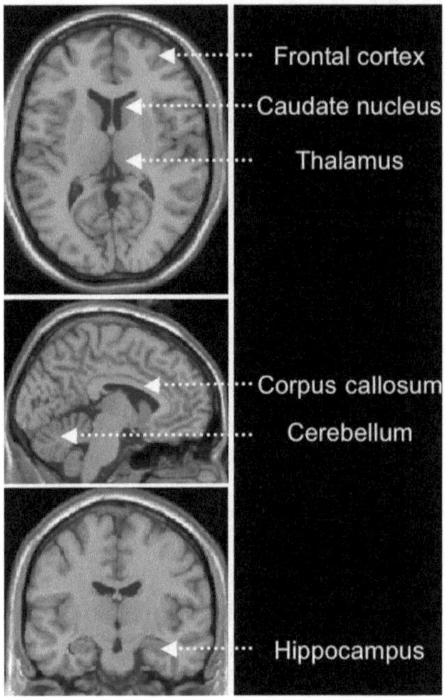

Fig. 1.6 Magnetic resonance images of the human brain. *Top*, axial section; *middle*, sagittal section; *bottom*, coronal section. From Paus (2009)

cannot visualize cyto- or chemoarchitecture with MRI, we cannot reveal cell-to-cell anatomical connections. (some of these limitations will be discussed in Chap. 9).

Finally, inter-individual variations in the state of brain structure and function assessed at a single time point are likely to reflect both *developmental processes* (that took place during the early periods of the individual's life) and *experience-based plasticity* (that may have a more recent origin). For example, overall brain size is likely to be the outcome of genetic and environmental influences acting in utero and during the first two years of post-natal life. This is simply related to the fact that brain growth is most dynamic during these periods: the human brain reaches ~420 cm^3 in volume (~36 % of adult values) at birth, 855 cm^3 (72 % of adult) by the end of the first post-natal year and 983 cm^3 (83 % of adult) at the end of the second year (Knickmeyer et al. 2008). On the other hand, more subtle inter-individual variations in brain structure (e.g. local volume in cortical grey matter) may reflect the individual's experiences. For example, we know that practicing a task like juggling every day will result in small, volumetric increases in the grey matter located in a cortical region important for processing visual motion. Thus, we observe cumulative functional engagement of this region being translated into a structural change (Draganski et al. 2004; Fig. 7.4). Clearly, when studying forces shaping the (healthy) human brain, we need to consider possible developmental timelines *and* cumulative exposures.

To summarize, when constructing a phenome of the human brain, we need to appreciate both the cellular and regional heterogeneity of this organ. When interpreting inter-individual variations in the brain phenotypes vis-à-vis those in genes and environment, we need to consider carefully the time period during which their "transactions" likely took place and the possible mechanisms mediating them (e.g. development- or experience-related plasticity).

1.4 Envirome and Phenome

An individual's entire genome can be had from a single blood sample. The same cannot be said for the individual's envirome and phenome. Furthermore, while the genome does not change over the lifespan, the envirome and phenome do. In the ideal world (for a scientist, anyway), one would sample all possible "exposures" with high frequency—say, on a daily basis—to reconstruct a multidimensional envirome of each day. From these daily snapshots, one would also create a historical record capturing an individual's cumulative exposures in different periods of his/her life. One would measure both the physical environment surrounding the person (e.g. air and water quality, noise levels) and his/her social environment (family, school, workplace and neighbourhood). And one would inquire about the person's life habits as a probabilistic indicator of "self-induced" exposures, related, for example, to diet or stress. (in Chap. 3, we will review the various types of environment and tools one may use to estimate them.)

In the same ideal world, one would pair up sampling of the environment with assessing the individual's phenome. Again, this would be done in a multidimensional fashion: we would not only measure the various properties of the brain but add measurements of other organs likely to impact the brain, such as the state of the cardiovascular and metabolic system. The individual's phenome would be characterized at multiple levels, from molecules (e.g. proteome, metabolome), through cells (transcriptome) and tissues (a culture of human-induced neuronal cells), to organs and systems. And this would be all done at least a few times during each of the more dynamic developmental periods (infancy, childhood, adolescence) and then in regular intervals—say, every 5 years—for the rest of the person's life.

Of course, we cannot assess the individual's envirome in the way outlined above. That being said, we might be able to obtain reasonable estimates of the various (cumulative) exposures (e.g. stressful life events, number of cigarettes smoked in the lifetime) by combining different sources of information at both individual and aggregate levels and by expressing exposures in probabilistic terms. To increase the accuracy of these estimates, we might want to "triangulate" exposures from multiple lines of evidence: (1) self-reports by the primary and secondary informants (e.g. family members), (2) personal records (e.g. from schools and from health-related databases) and (3) aggregate statistics (e.g. average income for a given profession or neighbourhood).

As for the phenome, the main challenge is the person's time. This is why technologies such as MRI are such powerful tools for acquiring multidimensional, system-level, quantitative phenotypes in large number of individuals (>10,000).

The efficiency of MRI—in particular, its widespread availability and non-invasiveness—makes it an ideal tool with which to answer the call for high-throughput and high-dimensional phenotyping (Houle et al. 2010).

Thus, as suggested by many, enviromics and phenomics are the main challenges of the post-genomic era (e.g. Anthony 2001, Freimer and Sabatti 2003; Bilder et al. 2009). State-of-the-art tools, enabling proper and thorough evaluation of exposures and system-level outcomes, must complement those available today in genomics, epigenomics and molecular phenomics.

1.5 Developmental Cascades

As outlined above, human brains and bodies comprise a set of complex systems (phenome) that interact with each other while being exposed to the influences of internal and external environments (envirome) over time. Clearly, an understanding of the interactions and feedback loops between these "spaces" (Fig. 1.7) is critical for achieving our goal, namely explaining inter-individual variability in the structural and functional state of the human brain.

While working towards this goal, we may consider taking *integrative* and *developmental* approaches to the study of processes leading to a particular state of the individual's brain predicting, in turn, his/her risk for developing a brain disorder. This approach resembles the concept of developmental cascades (Masten and Cicchetti 2010). This concept acknowledges complexity in time and space, where the former is real and the latter consists of constructs attempting to capture key elements of a given phenomenon at a particular level of analysis. Studies of developmental cascades require longitudinal data. Typically, the time dimension spans a few developmental periods (e.g. childhood, adolescence and young adulthood). The space factors must be multidimensional: constructs from various domains (e.g. cognition, mental and physical health, social environment) and levels (molecules, organs, individual behaviours and social relationships) must be sampled at each time point.

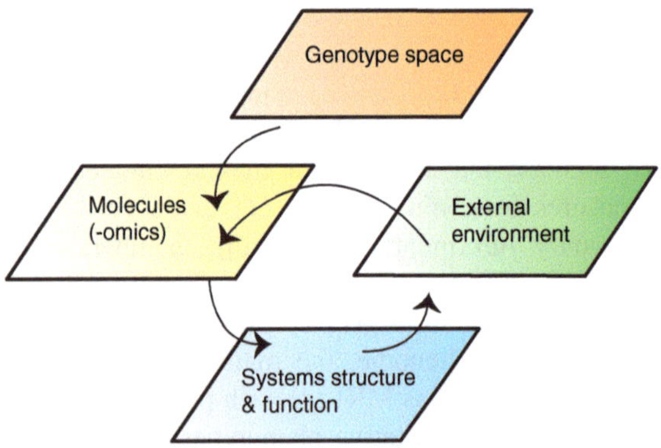

Fig. 1.7 The interacting spaces: genome, phenome (molecules and systems) and envirome. Adapted from Houle et al. (2010)

The crucial feature in a developmental cascade is a transaction between constructs across time. For example, does inter-individual variability in a particular cognitive process at Time 1 predict a particular aspect of mental health at Time 2 and, in turn, physical health at Time 4? Provided that enough data have been acquired with an appropriate sampling frequency, statistical models used to test developmental cascades can examine the directionality of such transactions, both across time (as noted above) and across space (i.e. across domains and/or levels; see Text Box 1.1.).

Text Box 1.1. Upward and downward developmental cascades

Pharmacological intervention may initiate an "upward" cascade where a drug—say, Ritalin—influences the molecular processes involved in particular neurotransmitter systems. These changes affect functioning of specific neural networks, and these, in turn, lead to behavioural changes of the child, impacting social dynamics in his/her class at school (Cicchetti and Curtis 2007). On the other hand, a "downward" cascade may begin with the child's maltreatment, lead to a repeated activation of the stress axis and eventually induce epigenetic changes in the relevant signalling pathway (e.g. methylation of glucocorticoid receptors). All this will result in reduced expression of the relevant genes. This change would, in turn, initiate an "upward" cascade, starting with a dysregulation of the stress system and ending with (internalizing) psychopathology (reviewed in Masten and Cicchetti 2010).

To repeat the obvious, the most exciting and challenging aspect of developmental cascades is the unfolding of transactions between constructs over time. The concept of developmental cascades illustrates nicely both the challenge and promise of an integrative, multi-level and developmental approach to the study of processes that lead to a particular *state* of brain structure and function. And, as indicated in the Preface, gaining insight into a process brings us closer to *prediction,* thus laying down the foundations for personalized preventive medicine.

1.6 Personalized Preventive Medicine

Personalized medicine has emerged as a new paradigm, an attempt to tailor strategies to a specific patient based on a biological marker or "biomarker" (Hamburg and Collins 2010). In this context, biomarkers can be defined as any objectively measured characteristic that has strong predictive value. The predictions can lead to diagnosis (whether you have a disease) and/or prognosis (how the disease will progress); they can also indicate drug response or drug toxicity (David Wishart 2011). Most significant of all, biomarkers can also predict the *probability* of getting a disease, underscoring the tremendous need for personalized preventive medicine. In Sect. 10, we introduce *personalized medicine* (tailoring treatment to an individual) and *preventive medicine* (prevention at a population level) and discuss the need to bring these two together in the form of *personalized preventive medicine*. We argue that it will be possible to identify profiles, in the phenome and genome of a given individual, which will predict (with high accuracy) the risk of developing a disease. If so, one would be able to offer a personalized strategy for reducing this risk. It is likely that biomarker profiles with the highest predictive value will be the ones marking the key "transactions" leading from exposures to outcomes; individuals with a particular constellation of distal causes and ensuing developmental cascades will likely differ in the state of their brains at a particular time in their life and, therefore, in their risk for developing a brain disorder.

In summary, the goal of population neuroscience is twofold: (1) to gain new knowledge about sources of inter-individual variability, both genetic and environmental, in the human brain and (2) to use this information for effective individualized prediction of risk and, in turn, the development of personalized primary and secondary preventive strategies tailored to the at-risk individual.

References

Anthony JC (2001) The promise of psychiatric enviromics. The Br J psychiatry Suppl 40:s8–11

Bilder RM, Sabb FW, Cannon TD, London ED, Jentsch JD, Parker DS, Poldrack RA, Evans C, Freimer NB (2009) Phenomics: the systematic study of phenotypes on a genome-wide scale. Neuroscience 164 (1):30–42. doi:S0306-4522(09)00048-7 [pii] 10.1016/j.neuroscience.2009.01.027

Braitenberg V (2001) Brain size and number of neurons: an exercise in synthetic neuroanatomy. J Comput Neurosci 10(1):71–77. doi:10.1023/A:1008920127052

Cicchetti D, Curtis WJE (2007) A multilevel approach to resilience. Dev Psychopathol 19:627–955

David Wishart (2011) Wishart, David. Personal communication, December 3

Draganski B, Gaser C, Busch V, Schuierer G, Bogdahn U, May A (2004) Neuroplasticity: changes in grey matter induced by training. Nature 427(6972):311–312

Freimer N, Sabatti C (2003) The human phenome project. Nat Genet 34 (1):15–21. doi:10.1038/ng0503-15 ng0503-15 [pii]

Hamburg MA, Collins FS (2010) The path to personalized medicine. New Engl J Med 363(4):301–314

Houle D, Govindaraju DR, Omholt S (2010) Phenomics: the next challenge. Nat Rev Genet 11(12):855–866. doi:10.1038/nrg2897

Knickmeyer RC, Gouttard S, Kang C, Evans D, Wilber K, Smith JK, Hamer RM, Lin W, Gerig G, Gilmore JH (2008) A structural MRI study of human brain development from birth to 2 years. J Neurosci 28(47):12176–12182. doi:10.1523/JNEUROSCI.3479-08.2008

Masten AS, Cicchetti D (2010) Developmental cascades. Dev Psychopathol 22(3):491–495. doi:10.1017/S0954579410000222

Paus T (2009) Brain development. In: Lerner RM, Steinberg L (eds) Handbook of adolescent psychology. 3 edn. Wiley and Sons, pp 95–115

Perera F, Herbstman J (2011) Prenatal environmental exposures, epigenetics, and disease. Reprod Toxicol 31(3):363–373. doi:10.1016/j.reprotox.2010.12.055

Peters A, Palay SL, Webster Hide F (1976) The fine structure of the nervous system. Saunders, Philadelphia

Chapter 2
History of the Key Disciplines

In this chapter, we will learn about the three disciplines central to population neuroscience: epidemiology, genetics and cognitive neuroscience—their beginnings and subsequent developments.

2.1 Epidemiology

In 1855, John Snow proposed that cholera is transmitted by water (Snow 1855). He reached this conclusion by getting addresses of cholera fatalities during the 1854 London epidemic and comparing them with the distribution of water sources (based on Morabia 2001). Water was provided by three different water-supply companies in the city; only one took water from a less polluted area of the Thames. The General Register Office, headed by William Farr, who created a standardized way of collecting and classifying causes of death in England, provided the addresses. Based on this database, and using "quantitative reasoning", Snow made his inference almost 30 years before anyone discovered the disease's causal agent, *Vibrio cholerae* (by Koch in 1884; reviewed in Howard-Jones 1984).

In mid-nineteenth-century medicine, such use of the "numerical method" was still uncommon (Lilienfeld and Lilienfeld 1982; see Text Box 2.1.). Thus, basic statistics, record keeping and careful observations and analyses of possible associations between exposures and outcomes gave rise to the science of epidemiology.

Text Box 2.1. The numerical method

The numerical method—the use of averages to characterize a phenomenon at a group, as opposed to an individual, level—was introduced by Pierre Charles Alexander Louis in 1828. William Guy, Professor of Forensic Medicine, and later of Hygiene, at the King's College in London, and Snow's contemporary, had this to say about the lack of numerical method in medicine: "[I]t is most worthy of remark, that the student of the exact sciences, who is familiar with the use of the most certain instruments of

T. Paus, *Population Neuroscience*, DOI: 10.1007/978-3-642-36450-1_2,

calculation, has never hesitated to apply the numerical method… whilst the medical man, whose science seems most to need the application of such a method, and to offer abundant occasions for its employment, still doubts its efficacy and prefers the obscurity of general phrases to the clearness and precision of numbers." [(Guy 1839), as quoted in (Lilienfeld and Lilienfeld 1982)].

In the ensuing 100 years, epidemiology developed into a discipline concerned with both proximal causes of diseases, such as exposures to infectious agents, and their distal causes, such as living conditions associated with the transmission of infectious agents and host susceptibility. Quickly, basic knowledge created by epidemiology developed into a "translational" discipline, namely public health or hygiene (Text Box 2.2.).

Text Box 2.2. Hygiene

To quote William Guy again: "As hygiene deals with mankind, not one by one, but in masses, its scientific method can be no other than that numerical method so often confounded with its leading application—statistics. If this word now meant what it originally did… then hygiene would take rank among its leading subdivisions as applying the great State-policy of prevention to health and disease" [Guy (1870), as quoted in (Lilienfeld and Lilienfeld 1982)].

In the process, generations of epidemiologists developed rigorous methodological foundations to address some of the issues fundamental to epidemiological studies, including study design (case–control and cohort studies) and the related possibility of a selection bias (representativeness), and confounding causal inference. Zhang et al. (2004) provide an illuminating review of these and other key concepts, which were reflected in classical textbooks of epidemiology published between 1935 and 1986.

In the second half of the twentieth century, epidemiology began to shift its focus away from infectious diseases, as these ceased to be the main source of mortality and morbidity in developed countries. To a great extent, this decreasing interest in infectious agents on the part of epidemiologists was due to the success of hygiene (and sanitation), bacteriology (and antibiotics) and virology (and vaccination). One may argue that many of these achievements were built on the knowledge generated by epidemiology, which also supported their implementation and evaluation.

What had to be tackled next? For many epidemiologists of the 1960s and 1970s, *social* determinants of health became the new domain of interest. This could be seen as reflecting both the growing health impact of chronic disorders—such as cardiovascular diseases—on morbidity and mortality in developed countries and the conceptual return to the original goals of epidemiology, namely the search for distal causes of disorders ("living conditions" of the hygiene era). Departments of "Social Medicine", "Community Medicine" or "Preventive Medicine" began opening their doors (Vandenbroucke 1990). Figure 2.1a illustrates this trend (between 1966 and 2005) as reflected in the total number of publications indexed in MEDLINE under various disease-related outcomes; chronic outcomes (rather than infectious diseases) appear most frequently (Cohen et al. 2007). When cross-referenced with subject headings indicating social factors (i.e. "residence characteristics", "social environment", "social conditions", "social change", "social

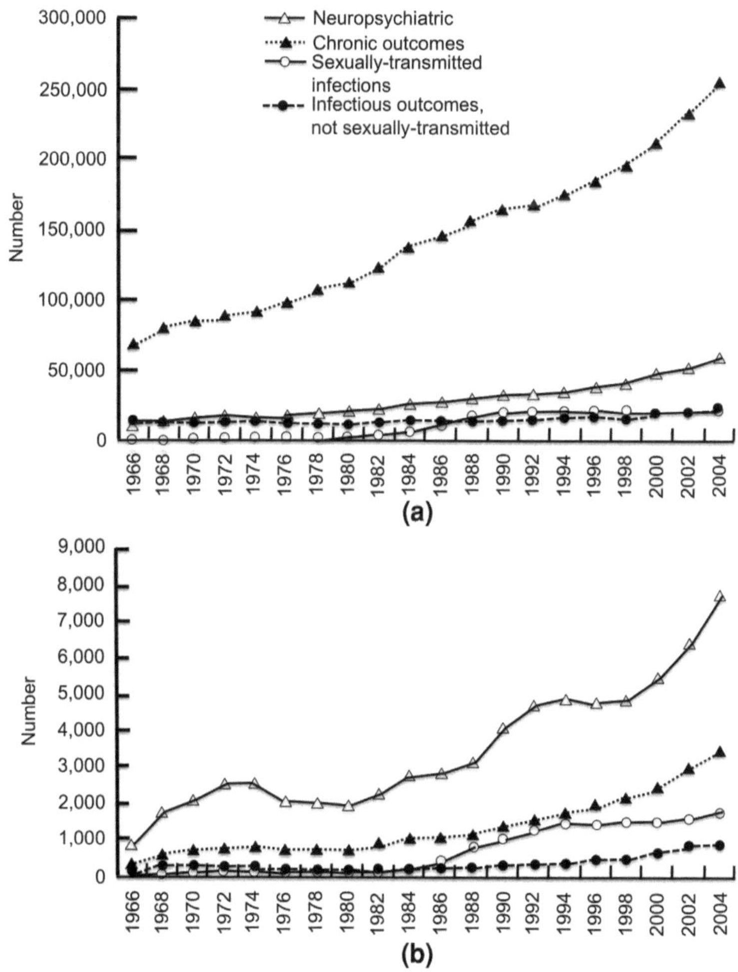

Fig. 2.1 (**a**) Number of citations per year indexed by Medline from 1966 to 2005 under subject headings related to important non-infectious, neuropsychiatric, infectious and sexually transmitted diseases. (**b**) Number of citations in the same disease categories additionally cross-referenced with subject headings related to social factors. From Cohen et al. (2007)

problems" and "social welfare"), psychiatric conditions and chronic disorders prevail (Fig. 2.1b). As we will note in (Sect. 10.1), many of these conditions are behind the growing gap between life expectancy (life span) and *healthy* life expectancy (health span). In the same analysis, Cohen et al. (2007) showed that the majority of review articles on social determinants and disease (published by November 2005) addressed cardiovascular diseases (28 %), cancer (15 %) and obesity (12 %).

The shift to social epidemiology and the work on socio-economic gradients of health, the role of social networks or the influence of built environments—to name but a few topics—represented a move away from "focusing on clinical diseases, one at a time" to the consideration of "psychosocial and cultural forces that compromise a person's ability to withstand insult" and establishing "the link between social forces and biologic processes" (Leonard Syme in Boyce 2011). The hope for preventive medicine represented the same "translational" aspiration of social epidemiology/medicine as that of the "hygienic" movement at the turn of the previous century.

In parallel with the emergence of social epidemiology, other strong trends were taking shape: molecular epidemiology (Kilbourne 1973; Schulte and Perera 1993) and genetic epidemiology (Morton et al. 1978; Khoury et al. 2004). The former returned primarily to the problem of infectious diseases, while the latter turned its attention to the study of genetic cases of non-infectious diseases, often the same ones studied by social epidemiologists. More recently, building on the successes of the Human Genome Project (HGP) (see Sect. 2.2), epidemiology joined forces with genomics to embark on "human genome epidemiology" (Khoury et al. 2004, 2010); the work discussed in the two editions of the Khoury et al. book is directly relevant to the topic of population neuroscience.

In summary, epidemiology traces its roots to the application of the "numerical method" (i.e. statistics) in the search for causes of epidemics and, in general, determinants of population health. From its initial interest in infectious diseases, it quickly expanded its scope to non-infectious diseases and, in the past 50 years, brought in tools and concepts from other disciplines, thus establishing new "hybrids", such as social epidemiology and genetic epidemiology.

2.2 Genetics

In 1866, Gregor Mendel formulated the laws of inheritance and introduced the concept of the allele as a fundamental unit of heredity (Mendel 1886). He used the "numerical method" to derive statistical rules explaining the pattern of observations made during his experimental work on plant hybridization; this work was carried out in a garden of an Augustinian monastery (Brno, Moravia, present-day Czech Republic) between 1856 and 1865. Unfortunately, Mendel's publication was missed by the mainstream science of the times, and his laws on inheritance were rediscovered only in 1900 by de Vries, Correns and von Tschermak. In the subsequent 50 years, a number of discoveries were made, laying down the biological foundations of genetics. Thus, in 1902, Walter Sutton suggested that "… the association of paternal and maternal chromosomes in pairs and the subsequent separation during the reducing division … may constitute the physical basis of the Mendelian law of heredity" (Sutton 1902, as quoted in Crow and Crow 2002).[1] The subsequent work established that "genes" do indeed reside on chromosomes (1910), that chromosomes contain DNA (1933) and that genes code for proteins (1941). Then, DNA was purified (1944), its first picture taken with X-ray diffraction (1952), and its 3D structure, a double helix, solved by Watson and Crick (1953; Fig. 2.2).

The next 20 years saw, among other discoveries, the cracking of the genetic code (1966)[2] and the isolation of restriction enzymes, a key discovery allowing the "cutting and pasting" of DNA. The following 20 years introduced two key methodological advancements, namely an electrophoresis-based method for DNA sequencing (1977) and polymerase chain reaction (1983) that enabled most of the work carried out in the 1990s by the HGP. The first disease-causing gene was identified (1989; cystic fibrosis trans-membrane conductance regulator), and the idea of doing genetic "fingerprinting" based on DNA polymorphisms was put forward (1989).

In 1990, the U.S. National Institutes of Health (NIH) and Department of Energy (DOE), together with their international partners, launched the HGP. Over the next 15 years, and with an annual budget of $ 200 million provided by the NIH and DOE, the HGP's plan was to map and sequence the 3.2-billion-nucleotide-long genome: creating genetic and physical maps of the human genome, identifying all genes and sequencing the entire DNA (Watson and Jordan 1989). The initial plan for the *genetic map* envisaged one containing ~1,000 highly informative markers, so-called microsatellites, placed—on average—with an interval of ~2–5 million nucleotides. The physical location of these markers, and genes, had to be specified in real "mile stones" placed on the genome's road: The plan called for the creation of a *physical map* to be based on the sequence-tagged sites[3]; a total of 30,000 of such sites were planned initially. The protein-coding genes were to be identified by synthesizing DNA complementary to messenger RNA, the complementary DNA or cDNA. The plan was to create a library of cDNA clones corresponding to all protein-coding genes in the human genome. At the time, the number of genes in the human genome was unknown, the estimates varying between 35,000 and 100,000 genes (Watson and Jordan 1989; Lander 2011). Finally, the goal of mapping the full DNA sequence of the human genome—that is, the sequence of its

---

[1] Students reading this text may find it interesting that Sutton made his discovery during his graduate studies, which he did not finish; he became a surgeon instead (Crow and Crow 2002).

[2] Unique triplets of messenger RNA nucleotides [i.e. codons] specify each of the 20 amino acids from which the proteins are built.

[3] Sequence-tagged sites (STSs) are short segments of DNA that occur only once in a genome.

MOLECULAR STRUCTURE OF NUCLEIC ACIDS

A Structure for Deoxyribose Nucleic Acid

WE wish to suggest a structure for the salt of deoxyribose nucleic acid (D.N.A.). This structure has novel features which are of considerable biological interest.

A structure for nucleic acid has already been proposed by Pauling and Corey[1]. They kindly made their manuscript available to us in advance of publication. Their model consists of three intertwined chains, with the phosphates near the fibre axis, and the bases on the outside. In our opinion, this structure is unsatisfactory for two reasons: (1) We believe that the material which gives the X-ray diagrams is the salt, not the free acid. Without the acidic hydrogen atoms it is not clear what forces would hold the structure together, especially as the negatively charged phosphates near the axis will repel each other. (2) Some of the van der Waals distances appear to be too small.

Another three-chain structure has also been suggested by Fraser (in the press). In his model the phosphates are on the outside and the bases on the inside, linked together by hydrogen bonds. This structure as described is rather ill-defined, and for this reason we shall not comment on it.

.

.

.

The previously published X-ray data[5,6] on deoxyribose nucleic acid are insufficient for a rigorous test of our structure. So far as we can tell, it is roughly compatible with the experimental data, but it must be regarded as unproved until it has been checked against more exact results. Some of these are given in the following communications. We were not aware of the details of the results presented there when we devised our structure, which rests mainly though not entirely on published experimental data and stereochemical arguments.

It has not escaped our notice that the specific pairing we have postulated immediately suggests a possible copying mechanism for the genetic material.

Full details of the structure, including the conditions assumed in building it, together with a set of co-ordinates for the atoms, will be published elsewhere.

Fig. 2.2 Two excerpts from the one-page report by Watson and Crick on DNA structure. From Watson and Crick (1953)

3.2 billion bases—by 2005 was clearly highly ambitious, especially given that the critical breakthroughs in parallel sequencing technology came only after the completion of the HGP (Lander 2011).

A quick look at Table 2.1 shows the impressive achievements of the HGP over its first eight years and its goals for the final five years (Collins et al. 1998). The

Table 2.1 Goals and the early achievements of the human genome project (1998)

Area	Goal 1993–1998	Status as of October 1998	Goals 1998–2003
Genetic map	Average 2–5 cm resolution	1 cm map published September 1994	Completed
Physical map	Map 30,000 STSs	52,000 STSs mapped	Completed
DNA sequence	Complete 80 Mb for all organisms by 1998	180 Mb human plus 111 Mb non-human	• Finish 1/3 of human sequence by end of 2001 • Working draft of remainder by end of 2001 • Complete human sequence by end of 2003
Sequencing technology	Evolutionary improvements and innovative technologies	90 Mb/year capacity at ~$0.50 per base Capillary array electrophoresis validated microfabrication feasible	Integrate and automate to achieve 500 Mb/year at ~$0.25 per base support innovation
Human sequence variation	Not a goal	–	100,000 mapped SNPs Develop technology
Gene identification	Develop technology	30,000 ESTs mapped	Full-length cDNAs
Functional analysis	Not a goal	–	Develop genomic-scale technologies
	E. coli complete sequence	Published September 1997	–
	Yeast complete sequence	Released April 1996	–
	C. elegans most of sequence	80 % complete	Complete December 1998
	Drosophilia begin sequencing	9 % done	Sequence by 2002
Model organisms	Mouse: map 10,000 STSs	12,000 STSs mapped	• Develop extensive genomic resources • Lay basis for finishing sequence by 2005 • Produce working draft before 2005

From Collins et al. (1998)

HGP finished in 2003, two years ahead of schedule. Through the HGP, we learned that the total number of protein-coding genes is much lower than expected (~21,000). The genetic map contained ~3,000 markers in 1994 but, in their 1998 report, Collins and colleagues planned to cover the genome with 100,000 single nucleotide polymorphisms (SNPs) over the next five years.[4] In the same report, Collins et al. (1998) predicted that "SNPs will be a boon for mapping complex traits such as cancer, diabetes, and mental illness", "... make possible genome-wide association studies..." and "... permit prediction of individual differences in drug response".

Since the completion of the HGP, a number of other genome-mapping efforts continue. For example, the International HapMap Project builds maps based on the combination of adjacent SNPs inherited together (haplotypes) and identifies SNPs that "tag" unique haplotypes; it is estimated that the human genome contains about 300,000–600,000 such haplotype-tagging SNPs, when compared with a total of 10 million common SNPs (hapmap.ncbi.nlm.nih.gov/thehapmap.html.en). Other efforts, enabled by the shift of parallel sequencing from the electrophoresis based to optical imaging based, focus on genome-wide epigenomic mapping of variations in chromatic modifications and methylation (Lander 2011).

In summary, genetics started in Mendel's garden with the application of statistics, continued through the basic discoveries explaining the molecular and cellular mechanisms of the genetic machinery, and culminated in the mapping of human genome sequences and the cataloguing of inter-individual variations in its various features. The impressive achievements of the HGP, vis-à-vis basic knowledge about the human genome, advancements of genotyping technology and the creation of genetic databases: all set the stage for mapping the sources of inter-individual variability of complex traits, including the structural and functional properties of the human brain.

2.3 Cognitive Neuroscience

In 1861, Broca suggested in his report to the (French) Anatomical Society that the third frontal convolution of the left frontal lobe is the "seat" of "articulated" (spoken) language (Broca 1861). He reached this conclusion by evaluating the symptoms, and dissecting the brain, of a patient called *Tan*, a 50-year-old man who died in his (surgical) care, due to gangrene of a leg. The patient had been admitted to the Bicêtre Hospital 21 years previously, a few months after losing the ability to speak—since then, the only words he could utter were "tan, tan" (Text Box 2.3.). Broca performed an autopsy within 24 h of *Tan*'s death; he concluded that the

[4] Today, over 10 million SNPs have been mapped.

centre of the pathology was in the posterior part of the third convolution of the left frontal lobe and suggested that this was the site of the original lesion that led *Tan* to lose his ability to speak, 21 years earlier (Broca 1861).

Text Box 2.3. The Patient Tan

Tan—a patient of Broca—was admitted to the Bicêtre Hospital a few months after losing the ability to speak. Reviewing medical records 21 years later (when he was treating Tan for gangrene), Broca put together a picture of a man who had been healthy and intelligent when originally admitted to the hospital, "who differed from a sane man only in the loss of articulated speech", and whose brain pathology developed over the ensuing 21 years, affecting next the movement of his right arm then the right leg, eventually leaving him bedridden. But even on his death bed, "*Tan* understood almost everything that was said to him… Numerical responses were those that he could make the best, by opening or closing his fingers. I asked him many time how many days he had been sick? He [sic] responded sometimes five days, sometimes six days. For how many years had he been at Bicêtre? He [sic] opened his hand four times in sequence, and then pointed with a single finger; this would make 21 years, and one saw above that this information was perfectly exact" (Broca 1861).

In marked contrast to epidemiology and genetics, the science of the brain began not with statistics but from mapping "symptoms" onto anatomical features of the brain, and it continued like that for the next 50 years. These maps were often created through similar studies of single patients using the so-called "anatomo–functional" correlation. The prevailing concept in Broca's day was that of localization of function. Various permutations of this concept—shaped both by its proponents and opponents—have continued to guide the empirical work of those interested in the relationship between the human brain and behaviour. By definition, this work has two components: the brain and the behaviour. In the course of the following 150 years, new tools and concepts emerged that refined the knowledge of the structural organization of the human brain on the one hand and the understanding of cognitive processes on the other hand.

Broca was a surgeon and an anatomist. In his report, he lamented: "One is left to be dominated by the old prejudice that the cerebral convolutions have nothing fixed about them, that they are simple folds made haphazardly, comparable to the disordered bendings of the intestinal loops" (Broca 1861). He was indeed very careful in describing the location of the lesion in *Tan*'s brain relative to the frontal convolutions (gyri and sulci). Anatomists of the late 19th and early 20th centuries spent a great deal of time mapping and naming the cerebral folds, studying their emergence during foetal development, as well as asking questions about possible relationships between the morphogenesis of the folds and psychiatric disorders (Marshall and Magoun 1998). In recent history (see below), some of this early anatomical work has been revived through in vivo imaging studies of the human brain. Thus, gross anatomy of the human brain has been used both as a road map for the functional localization and as a window into the brain development and its disorders.

The next stage of mapping brain anatomy moved from macroscopic to microscopic level. The centre of these efforts was the Kaiser Wilhelm Institute for Brain Research, founded by Oscar and Cecile Vogt in Berlin, Germany, in 1914. In the Vogt Neurobiologisches Laboratorium,[5] Korbinian Brodmann examined under the microscope the pattern of cellular architecture (cytoarchitecture) of the human cerebral cortex, identifying 43 distinct cytoarchitectonic areas (Brodmann 1909; Zilles and Amunts 2010). His teachers, Oskar and Cecile Vogt, used a different approach and arrived at 200 cortical areas, which differed in their pattern of myelination or myeloarchitecture (Vogt and Vogt 1919; Zilles and Amunts 2010). Today, we tend to view this work mostly as "atlas building" (reviewed in Toga et al. 2006), but its initiators were looking for (cellular) clues of functional specialization in the human brain—not unlike the mappers of the human genome searching in the DNA sequence for (molecular) clues of disorders and complex traits.

How could one examine functional specialization of the human brain without waiting for brain pathology to develop and for the patient to die in order to localize it? In 1934, a neurosurgeon called Penfield founded the Montreal Neurological Institute. There, he and his colleagues treated patients suffering from intractable epilepsy (Penfield 1977). The surgical treatment for epilepsy, which Penfield pioneered, based on a similar work done by Foerster in Breslau[6] in the 1920s, provided unique opportunities for studies of functional specialization. First, many brain surgeries required electrical stimulation of (or recordings from) the exposed cortex in order to map out the "eloquent" cortex (to be left untouched by the surgeon) and/or the epileptogenic cortex to be identified and, subsequently, removed. In this way, Penfield and his colleagues revealed, for example, the somatotopic organization of movement (motor "homunculus") and touch (sensory "homunculus"), and the exact extent and function of the Broca's area of the left frontal lobe (Penfield and Rasmussen 1950; Penfield and Jasper 1954). Second, the location and extent of each surgical removal were carefully documented in a drawing that, in turn, provided the necessary information for "anatomo-functional" correlations obtained in subsequent neuropsychological assessments of the (living) patient. Brenda Milner and her students reviewed such drawings and other relevant information, and classified the patients based on the site and side of the removal. Using precisely designed psychological tests, they characterized the type of (subtle) cognitive deficits associated with particular removals of brain tissue. Using this approach, Milner and her colleagues established, for example, the role

[5] The Laboratorium served as a nucleus for the development of the Kaiser Wilhelm Institute for Brain Research.

[6] Wroclaw, in the present-day Poland.

of the medial temporal lobes in the different types of (declarative) memory, and of the frontal lobes in "executive" functions and language (Milner 1998). It is important to note that, being an experimental psychologist by training, Milner brought her knowledge of (cognitive) psychology and the rigour of experimental design to these studies, which she and her group carried out for over 50 years, thus complementing the opportunities afforded by the neurosurgeons and their patients (Text Box 2.4.).

Text Box 2.4. An interview with Brenda Milner

Brenda is a well-known scientist who lives and works in Montreal. She does psychology. Brenda was born in 15 July 1918, in Manchester, England. Her favourite subjects in school were algebra and Latin. When she was a little girl, she also learned German. Brenda's father was 58 when she was born. Brenda did not go to school for she was taught by her father until she was 8 years old, that is when her father died. Brenda went to Cambridge University when she was 18. She was then very interested in maths but also in science. In those days, you had to decide what you want to do very early. So, she had to make an early choice whether to do maths or psychology. She picked psychology because she thought she would never be as good at maths as she was in science. Then, she got a scholarship to do psychology in England. But just then the war broke out. So, she then had to work at a radar station to find out what is the best way to design the radars. She then met her future husband Peter Milner. Peter was an engineer making the radars. Brenda was then sent with Peter to Montreal to work on an atomic energy project. Peter and Brenda quickly got married, packed, went by train to Scotland where they boarded a ferry called Queen Elizabeth and sailed to Boston. From Boston they went to Montreal. That was in October 1944. Brenda got a full time job as a professor at University of Montreal, teaching psychology. Peter went to Chaulk River to go on with his research on the atomic energy project.

Few years later, Brenda started going to interesting classes on psychology by Hebb at McGill. Brenda became very interested in what Hebb talked about. In the meanwhile, a brain surgeon Dr. Penfield was thinking about taking in one student of Hebb to test his patients. Hebb asked Brenda and she was very honoured and decided to go to the MNI.

Dr. Penfield was treating patients who had epilepsy by removing a piece of brain called hippocampus. But in one case, the patient lost his memory after the surgery. This surprised Dr. Penfield greatly because he had done this surgery so many times and yet it had never damaged the memory. Dr. Penfield then had another patient and it happened again. That frightened him! Brenda had to find out what caused it. She figured out that hippocampus is important for creating memories like what you ate today, what game you played and what you learned in class. This memory is called declarative.

Dr. Penfield and Brenda wrote a paper about what had been happening with his patients. In Hartford, Dr. Scoville read this paper and became soon interested because something similar happened to one of his patients, H.M. When he was 23, Scoville treated his epilepsy by taking out the hippocampus on both sides of the brain. After surgery, H.M lost declarative memory. He could still remember things from before the surgery because it was saved on the surface of the brain. If you would tell H.M your name and who you are and then go away for 5 min, H.M would not remember you. Brenda went to study H.M to learn more about the hippocampus and declarative memory. She tried out different memory tests with him.
Unfortunately, H.M could not go to her to Montreal all the way from Hartford so instead Brenda made the trip. Brenda had done a lot of tests with him but realized she wanted something new. So, she thought she would try the mirror drawing with him. In this test, you ask the patient to trace the outline of a star while looking at what you are doing only in a mirror. At first, you would do terribly but with practice it gets better. She tried the mirror drawing because it was something that was not used a lot and because it was easy to carry all the way to Hartford. When she first tried it with him he started out the same way as any of us would. But then later on she discovered something very amazing. Even though he did not remember he had been doing the mirror drawing yesterday she saw that he was getting better and better. Brenda Milner had proved that hippocampus is not important for all kinds of memory. What H.M could still do was to learn new skills. Like riding a bike, playing basketball, skating, skiing, tennis and mirror drawing! But he could never remember learning those skills. For that, he needed his hippocampus.
You see that I have not been using H.M's full name. That is because scientists do not want to harm the patients' feelings by being known as the patient who lost his memory, etc.
For Brenda Milner, her most interesting time at the MNI so far was in the 1950s. That was when most of scientific discoveries in psychology were made. At the end of my interview with Brenda, I asked her what the best thing about being a scientist is. She said that if you look at poetry, music or story writing, you will notice that it can be all as good a hundred years ago as it is now. But in science, the discoveries made now are always more advanced and better that the ones a hundred years ago. Science is always new and is meant to make our lives more interesting and understandable.
By Veronika Pausová (age 11), Montreal, 6 May 1998

Naturally, the work carried out in (human) patients was not occurring in a vacuum. Perhaps, the most relevant complementary knowledge was generated in non-human primates, mostly macaque monkeys, throughout the 1970s and 1980s. First of all, experiments carried out in monkeys allowed investigators to evaluate the behavioural impact of well-circumscribed *bilateral* lesions, thus "knocking off" the entire structural unit rather than only its half (as common in human studies). These experiments also provided a way of establishing which of the structures affected in patients are critical for a particular deficit; in the monkey, small structural "units" were lesioned separately or in combination with other brain regions (e.g. amygdala and hippocampus; Mishkin 1978). Second, new techniques were introduce to study neural connectivity; tracers such as horseradish peroxidase and, later, tritiated amino acids, were injected into distinct (cortical) regions to describe the pattern of afferent and efferent connections with other parts of the brain (e.g. Pandya and Kuypers 1969). Over the years, this kind of work allowed neuroscientists to draw, for example, a "wiring" diagram of the visual cortex (Felleman and van Essen 1991) and to create databases of cortico-cortical connectivity in the macaque brain (Kötter 2004). Third, neurophysiologists developed techniques for recording the electrical activity of single neurons ("single-unit" activity) in a behaving monkey. With this tool in hand, they went on to describe various features "coded" by brain cells throughout the cerebral cortex—from the orientation of visual stimuli in the primary visual cortex (Hubel and Wiesel 1968), through rotation of the limbs at their joints in the parietal cortex (Mountcastle 1975) to size, shape and colour of visual stimuli in the temporal cortex (Gross 1969) and working memory in the prefrontal cortex (Fuster 1973). Together with the growing knowledge of brain connectivity, these studies forced neuroscientists to think about the brain as a collection of highly specialized but tightly interconnected brain regions.

In the 1980s, positron emission tomography (PET) brought an apparent revolution to the studies of brain–behaviour relationships in humans. Although PET (as we know it) was invented in the 1970s, it was not until the second half of the 1980s that neuroscientists started using PET for mapping cognitive processes in healthy human volunteers. To a large extent, this was due to a series of innovations that emerged from the group of Raichle at St. Louis Washington University (Text Box 2.5.).

Text Box 2.5. Key Innovations that Enabled the Mapping Brain Function with PET

The key innovations enabling the use of PET for studying brain–behaviour relations in healthy volunteers included: (1) the use of ^{15}O-labelled water for measuring cerebral blood flow as an index of brain activity; (2) introduction of the "subtraction" technique (Task A minus Task B) to isolate brain regions associated with a particular set of cognitive processes; and (3) statistical analysis of such differences between A and B in a group of individuals, by mapping their brains into the common 3D space, calculating a statistics for each voxel and reporting X, Y and Z coordinates of the significant voxels in a standardized stereotaxic space (Raichle 2009).

Over the next 20 years, this brain-mapping approach—and its expansion due to the development of non-invasive (radiation free) and more widely available functional magnetic resonance imaging (fMRI)—generated literally thousands of studies, reporting brain responses associated with various sensory, motor and cognitive processes. By virtue of recording task-related brain responses throughout the entire brain, functional neuroimaging both enabled and forced network-based analysis of brain function. Terms such as functional and effective connectivity were introduced to refer to different types of inter-regional relationships in the recorded signal (Friston 2002). At the same time, various forms of structural MRI opened up the opportunities for assessing structural properties of the grey and white matter, with sophisticated voxel-wise analysis of brain images carried out in an automatic "pipeline" manner. The number of participants scanned in any given study soared, from an average of seven participants per PET study (reviewed in Paus et al. 1998) to hundreds of participants enrolled in a given MRI-based population study (see Chap. 8). Perhaps not surprisingly, MRI-based imaging has become a tool of choice for high-throughput brain phenomics.

In summary, the study of brain–behaviour relationships in humans has proceeded through the mapping of clinical observations (symptoms) onto brain lesions, which were identified *post-mortem*, and cognitive test–based assessments onto the maps of surgical removals (or brain images) obtained in living patients, all the way to functional and structural neuroimaging studies of healthy individuals. The work in humans has been always informed by detailed studies of experimental animals, mostly macaque monkeys, which have provided the details of neural connectivity and revealed functional specialization of the primate brain at a single-cell level.

References

Boyce WT (2011) A conversation with Leonard Syme. Epidemiology 22:867–871

Broca P (1861) Remarks on the seat of the faculty of articulated language, following an observation of aphemia [loss of speech], vol 6. Bulletin de la Société Anatomique, Paris pp 330–35

Brodmann K (1909) Vergleichende Lokalisation der Grosshirnrinde in ihren Prinzipien dargestellt aufgrund des Zellenaufbaus. Barth, Leipzig

Cohen JM, Wilson ML, Aiello AE (2007) Analysis of social epidemiology research on infectious diseases: historical patterns and future opportunities. J Epidemiol Commun Health 61(12):1021–1027. doi:10.1136/jech.2006.057216

Collins FS, Patrinos A, Jordan E, Chakravarti A, Gesteland R, Walters L (1998) New goals for the U.S. Human Genome Project: 1998–2003. Science 282(5389):682–689

Crow EW, Crow JF (2002) 100 years ago: Walter Sutton and the chromosome theory of heredity. Genetics 160(1):1–4

Felleman DJ, Van Essen DC (1991) Distributed hierarchical processing in the primate cerebral cortex. Cereb Cortex 1(1):1–47

Friston K (2002) Functional integration and inference in the brain. Prog Neurobiol 68(2):113–143

Fuster JM (1973) Unit activity in prefrontal cortex during delayed-response performance: neuronal correlates of transient memory. J Neurophysiol 36(1):61–78

Gross CG, Bender DB, Rocha-Miranda CE (1969) Visual receptive fields of neurons in inferotemporal cortex of the monkey. Science 166:1303–1306

Guy WA (1839) On the value of the numerical method as applied to science, but especially to physiology and medicine. J Roy Stat Soc 2:25–47

Guy WA (1870) Public health: a popular introduction to sanitary science, being a history of the prevalent and fatal diseases of the English population from the earliest times to the end of the eighteenth century. Renshaw, London

Hubel DH, Wiesel TN (1968) Receptive fields and functional architecture of monkey striate cortex. J Physiol 195(1):215–243

Khoury MJ, Little J, Burke W (eds) (2004) Human genome epidemiology. Edited by New York:Oxford University Press

Khoury MJ (2010) Human genome epidemiology 2 edn. Oxford University Press, Oxford

Kilbourne ED (1973) The molecular epidemiology of influenza. J Infect Dis 127(4):478–487

Howard-Jones N (1984) Robert Koch and the cholera vibrio: a centenary. British Med J 288:379–381

Kotter R (2004) Online retrieval, processing, and visualization of primate connectivity data from the CoCoMac database. Neuroinformatics 2(2):127–144. doi:10.1385/NI:2:2:127

Lander ES (2011) Initial impact of the sequencing of the human genome. Nature 470(7333): 187–197. doi:10.1038/nature09792

Lilienfeld AM, Lilienfeld DE (1982) Epidemiology and the public health movement: a historical perspective. J Pub Health Policy 3(2):140–149

Marshall LH, Magoun HW (1998) Discoveries in the human brain. Human Press Inc., Totowa

Mendel JG (1886) Versuche über pflanzen-hybriden (experiments on plan hybridization). Verhandlungen des Naturforschenden Vereins zu Brünn (Proceedings of the natural history society of Brno)

Milner B (1998) History of neuroscience in autobiography, vol 2. Academic Press, San Diego

Mishkin M (1978) Memory in monkeys severely impaired by combined but not by separate removal of amygdala and hippocampus. Nature 273(5660):297–298

Morabia A (2001) Snow and Farr: a scientific duet. Soz Praventivmed 46(4):223–224

Morton NE, Chung CS (eds) (1978) Genetic Epidemiology. Academic Press, New York

Mountcastle VB, Lynch JC, Georgopoulos A, Sakata H, Acuna C (1975) Posterior parietal association cortex of the monkey: command functions for operations within extra personal space. J Neurophysiol 38(4):871–908

Pandya DN, Kuypers HG (1969) Cortico-cortical connections in the rhesus monkey. Brain Res 13(1):13–36

Paus T, Koski L, Caramanos Z, Westbury C (1998) Regional differences in the effects of task difficulty and motor output on blood flow response in the human anterior cingulate cortex: a review of 107 PET activation studies. NeuroReport 9(9):R37–R47

Penfield W (1977) No man alone: a surgeon's life. Little, Brown and Co, Boston

Penfield W, Jasper HH (1954) Epilepsy and the functional anatomy of the human brain. Little, Brown, & Co, Boston

Penfield W, Rasmussen T (1950) The cerebral cortex of man: a clinical study of localization of function. Macmillan, New York

Raichle ME (2009) A brief history of human brain mapping. Trends Neurosci 32(2):118–126. doi:10.1016/j.tins.2008.11.001

Schulte PA, Perera FP (1993) Molecular epidemiology: principles and practice. Academic Press, San Diego

Snow J (1855) On the mode of communication of cholera 2nd ed. J Churchill, London

Sutton WS (1902) On the morphology of the chromosome group in Brachystola magna. Biol Bull 4:24–39

Toga AW, Thompson PM, Mori S, Amunts K, Zilles K (2006) Towards multimodal atlases of thehuman brain. Nat Rev Neurosci 7:952–966

Vandenbroucke JP (1990) Epidemiology in transition: a historical hypothesis. Epidemiology 1(2):164–167

Vogt CVO (1919) Allgemeinere ergebnisse unserer hirnforschung. J Psychol Neurol 25:292–398
Watson JD, Crick FHC (1953) A structure for deoxyribose nucleic acid. Nature 171:737–738
Watson JD, Jordan E (1989) The human genome program at the national institutes of health. Genomics 5:654–656
Zhang FF et al. (2004) Evolution of epidemiologic methods and concepts in selected textbooks of the 20th century. Soz Präventivmed 49:97–104
Zilles K, Amunts K (2010) Centenary of Brodmann's map–conception and fate. Nat Rev Neurosci 11(2):139–145. doi:10.1038/nrn2776

Chapter 3
Enviromics

In this chapter, we will explore the environment surrounding us, the envirome, and the tools one may employ to measure it, enviromics. We will (1) consider different levels of "space" (in and around us); (2) highlight its time dimension; and (3) briefly discuss conceptual issues related to possible interdependencies between the environment and our behaviour.

3.1 Environmental Space

As we explained in Sect. 1.4, we are both "recipients" and "creators" of our environment. That is to say, the environment effects change in us and we also effect changes in the environment.

We can—albeit somewhat arbitrarily—divide environment into external and internal. The external can be observed from the outside (and/or self-reported). The internal is part of the body, so we require special tools to assess it. The external environment exists in both physical (e.g. air pollution, noise) space and social (family, neighbourhood) space. The internal environment is found on various body surfaces (e.g. microbes on our skin and in the gut) and, of course, in the blood (e.g. hormones, micronutrients). Given our interest in explaining sources of inter-individual variations, we need to consider and sample environment from the perspective of each individual person (Fig. 3.1).

Because the cells of various tissues are exposed to hundreds of small molecules (metabolites), hormones and enzymes circulating through the body, the *internal environment* of an individual is perhaps best characterized by the (chemical) content of his/her blood. Most of those molecules are able to cross the blood–brain barrier, so they also bathe the brain cells. (we will describe tools for measuring the levels of such molecules [metabolomics and proteomics] in Chap. 6 and touch on the issues of their dependence/independence on the individual's behaviour in Sect. 3.3 of this chapter.)

A major (external) source of small molecules is the food we eat. A number of instruments have been designed to capture the consumption frequency of a wide

T. Paus, *Population Neuroscience*, DOI: 10.1007/978-3-642-36450-1_3,

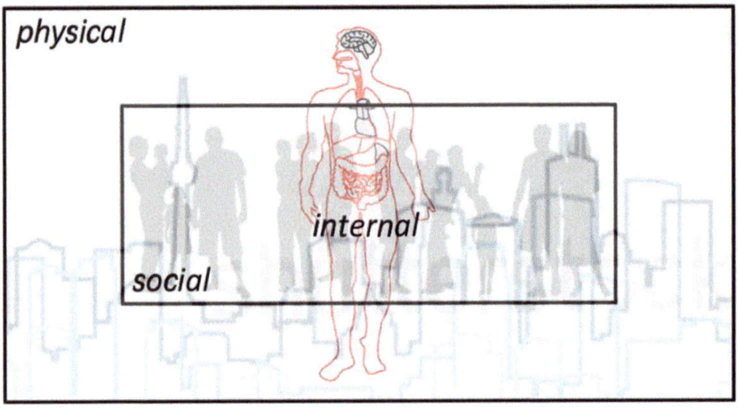

Fig. 3.1 An individual and his/her internal, social and physical environments

range of foods and drinks. In the ALSPAC Food Frequency Questionnaire, for example, informants are asked to indicate how often a given food/drink is "typically" consumed (Rogers and Emmett 1998; Northstone et al. 2011). A 24-hour food-recall samples in more detail all food consumed over the previous 24 h through a structured interview of the individual by a dietician; this information is then analysed using a recipe file to estimate the intake of macronutrients and micronutrients (e.g. Slimani et al. 2007).

The *"body" environment* represents the body's interface between internal and external environments and consists of the epithelial surfaces lining the gastrointestinal, urinary and respiratory tracts and the epidermal surface of our skin. These surfaces contain a kind of auxiliary biological system called the microbiome, which coexists with each human being (Text Box 3.1.). This intimate living arrangement with trillions of other organisms has profound effects on the internal environment and therefore on human health. For example, microbial flora of the gut exert(s) a considerable influence on the absorption of nutrients, such as vitamins and essential fatty acids (James and Garza 2012); in turn, nutrients affect multiple biological processes occurring every day in the brain—from the synthesis of neurotransmitters, through the reshaping of dendrites, to the myelination of axons.

Text Box 3.1. Microbiome

Comprised mostly of bacteria and other prokaryotes, the microbiome's total number of cells can exceed that of the human host by several orders of magnitude. In fact, 100 trillion cells come from more than 1,000 species (Tremaroli and Bäckhed 2012). Is it possible to characterize an individual's "body" environment? Thanks to detailed knowledge of the microbial genome, we can reconstruct microbial flora present in the samples of faeces and urine, as well as in the cheek and skin swabs and, in turn, model the likely environment created by the flora together with its possible impact on the relevant biological processes (Kuczynski et al. 2011).

The social environment reflects the complexity of who we are as social beings. Its health implications are indisputable. It is a well-known fact that social isolation predicts morbidity and mortality (House et al. 1988; Holt-Lunstad et al. 2010). Put simply, "humans need others to survive and prosper" (Cacioppo et al. 2011). Expanding circles of the social space includes the individual's family, workplace, neighbourhood, community and culture. Without a doubt, these social environments play a key role in shaping our brains and behaviour. But as discussed in Sect. 3.3, social environments in general, and the family environment in particular, are often confounded by our genes. We need to keep this in mind when using various measures of social environment as "exposures" when, in fact, they might represent "outcomes".

When reconstructing a child's family environment, we need to focus on the primary caregivers, their circumstances and their behaviour. To start with, socio-economic circumstances in childhood can foreshadow adult health (Text Box 3.2.). It is relatively easy to acquire information about basic socio-demographic characteristics of the child's family—indeed, a number of instruments exist that serve this purpose well (Table 3.1).

Text Box 3.2. Socio-economic circumstances in childhood predict adult health

In a prospective study, 1,131 male graduates of the Johns Hopkins University School of Medicine (graduating years 1948–1964) were assessed during medical school, asked about the occupation of their fathers, then followed up for up to 40 years with questionnaires. Low (vs. high) socio-economic status (SES) during childhood (e.g. father being a farmer vs. a professional) was associated with higher incidence of coronary heart disease before the age of 50 years, "despite physicians' high level of SES as adults, their medical knowledge, and their access to high-quality health care" (Kittleson et al. 2006).

Maternal care is a powerful modulator of the brain systems important for stress responsivity (reviewed in Champagne and Curley 2009). Therefore, it is vital to examine caregiving-related issues—from ascertaining the primary caregiver to evaluating family relationships in infancy and childhood. A number of instruments have been designed and validated for this purpose (Table 3.1). In general, information collected with such instruments should be corroborated by multiple informants whenever possible.

During adolescence, the influence of peers and the school environment begins to combine with that of the nuclear family. A number of instruments are suitable

Table 3.1 Assessing social and physical environment

Environment	Domain	Tool/questionnaire
Social	Family	Socio-demographic questionnaire of the MacArthur network[a]
	Child care and parenting	ALSPAC child care and parenting questionnaire
		Life history calendar (1)[b]
		Parental bonding instrument (2)
		University of Wisconsin–family assessment Caregiver scale (3)
		Parenting stress index (4)
		Dyadic adjustment scale (5)
		Childhood trauma questionnaire (6)
	Adolescence and peers	Positive youth development
	Workplace	Socio-demographic questionnaire of the MacArthur network
		Perceived stress scale by Cohen
	Neighbourhood	PhenXToolkit for community cohesion[c]
	Community and culture	Social network index (7)
		Interpersonal support evaluation scale (8)
		MOS social support survey (9)
		PhenXToolkit—acculturation protocol[d]
Built	Neighbourhood amenities	PhenXToolkit for healthy food[e]
Physical	Lighting and noise	Aggregate
	Water, air, contaminants	Aggregate

[a] http://www.macses.ucsf.edu/research/socialenviron/sociodemographic.php
[b] (1) Freedman et al. 1988; (2) Parker 1990; (3) Greenberg et al. 1993; (4) Abidin 1989; (5) Spanier 1989; (6) Bernstein and Fink 1998; (7) Cohen and Willis 1985; (8) Cohen et al. 1985; (9) Sherbourne and Stewart 1991.
[c] https://www.phenxtoolkit.org/index.php?pageLink=browse.protocoldetails&id=210801
[d] https://www.phenxtoolkit.org/index.php?pageLink=browse.protocoldetails&id=180101
[e] https://www.phenxtoolkit.org/index.php?pageLink=browse.protocoldetails&id=210701

for capturing the nature of social experiences in this period of life. For example, Lerner and colleagues developed a questionnaire called Positive Youth Development ($\sim$300 items), based on the "5 Cs" model: competence, confidence, character, social connection and caring. The questionnaire inquires about "contextual assets" available to the adolescent vis-à-vis his/her friends, community, family and school climate, as well as his/her participation in extra-curricular activities (Lerner et al. 2005).

In adult life, "exposures" associated with the workplace environment represent a powerful *cumulative* factor shaping our brains, eight hours a day, five days a week, for 40+ years. The brains of taxi drivers (Maguire et al. 2000), musicians (Gaser and Schlaug 2003) or typists (Scheibel et al. 1990) differ from the "average brain" in certain aspects relevant to their profession, thus pointing to experience-driven plasticity as one of the sources of inter-individual variability in brain structure and function. But the workplace is also the source of other "exposures" related to social interactions, the most common one being stress related to job demands and social hierarchies (Text Box 3.3.). Thus, in addition to capturing the individual's occupation, employment history and income, we should also attempt to characterize the nature of interpersonal interactions at the workplace, as well as job satisfaction and perceived stress. For adults as well as children, family is an important source of social interactions: we should record marital status, number of children and the physical and psychological closeness of an extended family, as well as the frequency and context of family interactions (e.g. common meals).

Text Box 3.3. Gender and stress in a workplace

In an interesting study pertaining to the gender of "superordinates" (supervisors, managers) and "subordinates" (workers), over 1,500 workers were asked about the gender of their supervisors and about their own psychological (e.g. "felt anxious or tense") and physical (e.g. headaches) distress in the past seven days. For men, being supervised by one man and one woman was associated with less psychological and physical distress that having a single male supervisor. For women, having a female supervisor (whether a single one or as a member of a supervising pair) was associated with more distress, as compared to having a single male supervisor (Schieman and McMullen 2008).

Finally, the *community and culture* in which the person lives represent—from a social perspective—a potential source of both positive and negative "exposures"; by definition, these depend on the nature of social *interactions* between the individual and other people in his/her physical (or virtual) community. Here, broader social networks available to the individual can be assessed using a variety of instruments. Finally, cultural identity should be carefully examined, especially in population studies based in multicultural settings, where immigration-related factors provide powerful sources of inter-individual variations Text Box 3.4.

Text Box 3.4. Acculturation

In their classical study of Japanese immigrants to California, Marmot et al. (1975) discovered that immigrants who adopted a "Western culture" had the rate of coronary heart disease five times higher than that of immigrants who continued living in "traditional ways". This difference was independent of diet, serum cholesterol or smoking—suggesting a role for psychosocial factors like social support (Marmot et al. 1975; Marmot and Syme1976).

The built environment represents the infrastructure in which most of our life unfolds: family dwelling, school, workplace, local community, city or countryside. Each of them generates a particular social and physical environment that, in turn, can be viewed as a set of "exposures". There exist several self-report questionnaires that measure the effects of the built environment. They inquire about housing, amenities available in the informant's neighbourhood or town (e.g. shops, gyms, parks, health care), transportation and neighbourhood safety. Relevant questions about the size of an elementary school and high school or the type of workplace (size, possible environmental contaminants) can be asked when inquiring about education and occupation, respectively. In addition to the reports of individual informants, built environments can be also characterized at an aggregate level (see below).

The physical environment largely depends on the built environment and includes both electroacoustic (lighting, noise) and physical (water, air, environmental contaminants in buildings or earth) factors. The physical environment can be inferred from the individual's reports on where he/she has lived and worked. More precisely, albeit at an aggregate rather than at an individual level, the physical environment can be estimated from either geographically anchored measurements taken by various (governmental) agencies or in targeted investigations (Text Box 3.5).

Text Box 3.5. Air pollution and the learning brain

Researchers in Barcelona (Jordi Sunyer, Jesus Pujol and colleagues) are conducting a longitudinal study on air pollution in children, aged 9–10 years. They have selected 20 schools located in high-car-traffic areas with high air pollution and 19 control schools with low air pollution (well matched with other factors, such as the SES of the students' families). To assess exposures, they sample air both inside and outside each school using a variety of devices (e.g. aethalometer). These will allow the researchers to quantify the mass, number and chemical composition of particulate matter (up to $\leq$10 μm in aerodynamic diameter). To assess outcomes, the researchers measure cognitive abilities in a total of 2,800 children from the 39 schools. In a subsample of 400 children, they will also obtain structural and functional brain MRIs (Jordi Sunyer, personal communication, July 16, 2012).

As flagged above, data collected at *aggregate levels*—vis-à-vis social, built and especially physical environments—represent a unique opportunity for complementing self-reports as objectively as possible. Thus, indicators for social cohesion include measures such as voter turnout, recycling rates, newspaper readership, charitable donations and feelings of safety. Indicators of neighbourhood quality include information about general economic wealth (based on tax returns), neighbourhood type (e.g. ratio of private homes to businesses), rental costs and vacancy rates, shopping facilities, crime rates and vandalism. Indicators for environmental quality include green space per capita, air and water quality and climate. These aggregate indicators may be linked to geographical units such as census or postal code and, in turn, to the individual participant.

In the not too distant future, it is likely that new *technological developments* will be brought to bear on efficient and objective data gathering for social and physical environments. For example, at a workshop organized by the Stanford Centre for Innovations in Learning (Pea et al. 2004), workshop participants suggested various ideas for the use of sensors in data gathering (e.g. "sensor-augmented costume" (Fig. 3.2)). A combination of sensor-based technology and/or mobile phones (Kim et al. 2008), with computer-based analysis of collected signals/images, would provide a powerful tool for obtaining a series of snapshots of the individual's daily activities and environment. In terms of longitudinal data, it might be possible to make use of information gathered during the purchase of various goods, including food, so that one can build a long-term profile akin to customer profiling used by marketing companies (e.g. www.manifolddatamining.com). Appropriate modifications of such technologies and data handling would be mandatory, to ensure full protection of confidentiality. This would be an ethically acceptable use of the information gathered in this manner. Until such ethical issues are resolved, such approaches could be used at aggregate levels.

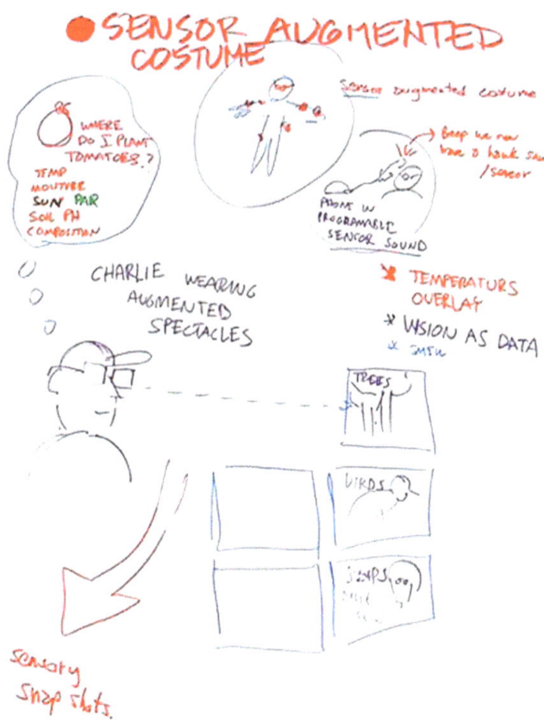

Fig. 3.2 Sensor-augmented costume. A drawing illustrating ideas for the use of sensors for data gathering, generated by a participant of a workshop organized by the Stanford centre for innovations in learning. Reprinted from Pea et al. (2004)

3.2 Environment in Time

The various environmental "spaces" introduced above operate over a long period of time, shaping the human brain and behaviour from conception onwards. To understand these forces, we need a historical record capturing the cumulative exposures of the individual in different periods of his/her life. Birth cohorts are in an enviable position to reconstruct such historical records, from data collected in a prospective fashion, including the most dynamic developmental periods—namely infancy, childhood and adolescence—when many "transactions" take place across levels and domains (see Sect. 1.5, on Developmental Cascades). Clearly, measuring the envirome and brain phenome concurrently over time would be ideal (Sect. 1.4). But to study a representative sample of individuals from cradle to grave is not possible for a number of reasons, including challenges associated with a life-term commitment of the participants, a probable *selection bias* as time progresses and the fact that continuous advancements in brain-mapping technology affects the tools available at any given time point of a decades-long study.

At least two practical solutions are available today. *First*, one can use "early" envirome—collected in a prospective fashion in a birth cohort—as a predictor of brain phenome acquired at some later time point in adulthood. *Second*, one can reconstruct an "early" envirome in a retrospective fashion using multiple informants, especially the parents, while acquiring the brain phenome at a later point in life. Needless to say, the time elapsed between the "early" and "later" periods would likely influence both accuracy and validity of the information collected in this way.

New knowledge and technological advancements may also bring alternative solutions. The environment induces epigenetic modifications in our genome (Sect. 1.2 and Chap. 5). As the mapping of the epigenome progresses, we might be able to get to a point when cumulative effects of (some) environmental exposures can be "read off" from the pattern of such modifications; the current work on maternal care and epigenetic modifications of the glucocorticoid receptor gene in experimental animals is of note here (Champagne and Curley 2009). Furthermore, having access to blood samples from different developmental periods of the individual's life would then allow us to estimate when a particular exposure took place. Finally, such an epigenetic approach may become critical for understanding trans-generational transmission of environmental exposures (see Chap. 5 for details).

3.3 Interdependencies: Environment, Behaviour and Genes

Many of our food choices are habitual and tend to echo the eating habits of family, friends and culture (including influences from advertising). Eating choices are also shaped by family budget and food availability (e.g. shops and seasonality of

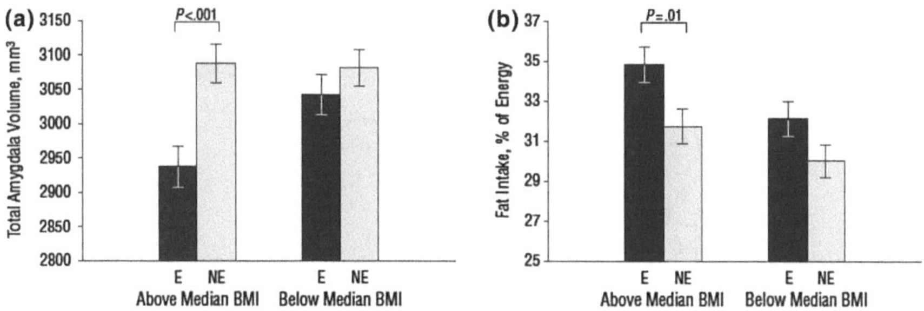

Fig. 3.3 Association of prenatal exposure to maternal cigarette smoking with amygdala volume (A) and fat intake (B) in individuals with above median and below median body mass index (*BMI*). The median split was performed separately in exposed (*E*) and non-exposed (*NE*) individuals on age-adjusted and sex-adjusted BMI. The data are presented as age-adjusted and sex-adjusted means and standard deviations. From Haghighi et al. (2013)

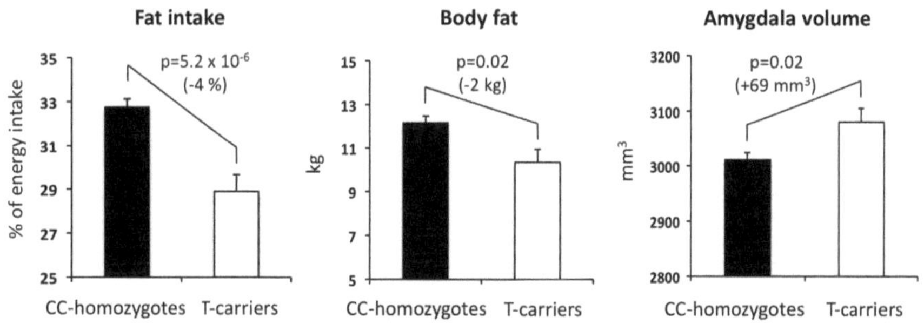

Fig. 3.4 Association of the opioid receptor mu 1 (*OPRM1*) locus (rs2281617) with fat preference, body fat and amygdala volume. Means ± standard errors, adjusted for age, sex and height (body fat only). From Haghighi et al. (2012b)

certain items). Biologically based food preferences play an additional role and could indicate early environmental influences (Haghighi, Pausova et al. 2013; Fig. 3.3) and/or specific genetic factors (Haghighi et al. 2012b; Fig. 3.4). In general, the envirome is both the source and target of inter-individual variability in brain and behaviour.

The complexity of causation—acting across domains and over time—cannot be addressed in individual studies. It might be approximated when targeting a particular area, either through dense longitudinal sampling of a number of related domains or through an experimental manipulation/intervention in one of the domains (Masten and Cicchetti 2010). Furthermore, environment may *correlate* with genes; hence, the crucial element of adoption (reared apart, reared together) in twin studies aimed at disambiguating the influence of genes and environment (Plomin 1994; Bouchard and McGue 2003). As see in Fig. 3.5, similarity between two traits—expressed as a correlation—can be attributed to common genes alone, albeit at a different "dose". In this context, the fact that child's vocabulary

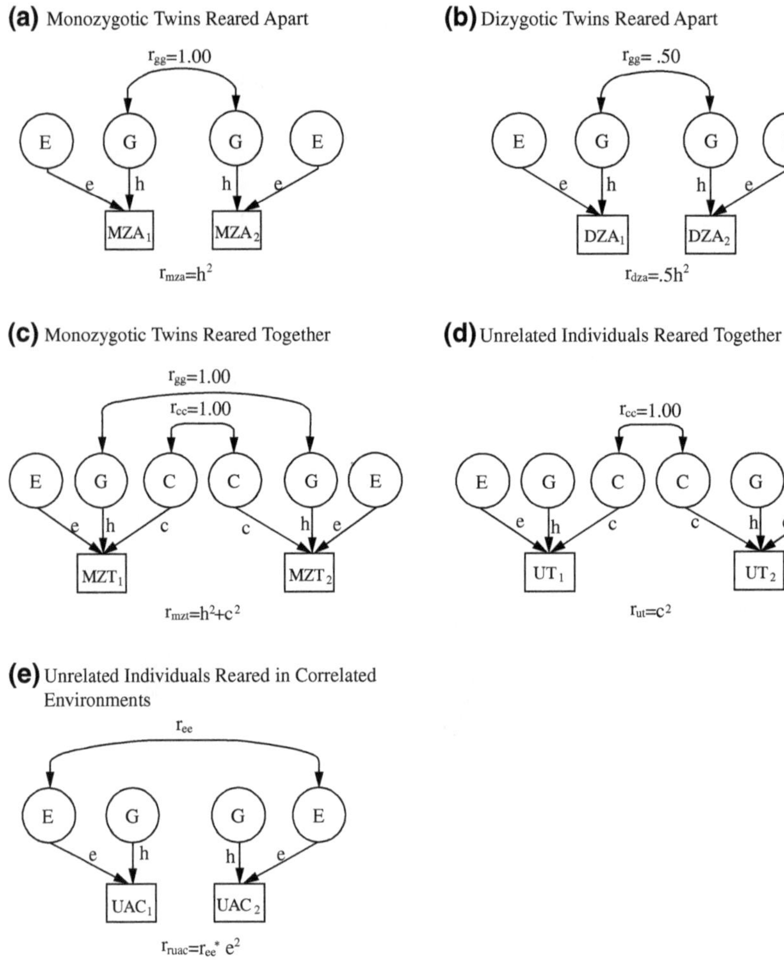

Fig. 3.5 Contributions of genes and environment to trait similarity (adapted from Bouchard and McGue 2003). G, genes; C, common (shared) environment; E, unique (non-shared) environment

correlates with the number of books in his/her parents' library likely reflects gene–environment correlation with the parents, which confounds an "exposure" by a transmission of genes related, perhaps, to general intelligence (Bouchard and McGue 2003).

To conclude, the envirome represents a critical piece of the puzzle of inter-individual variability in brain and behaviour. As illustrated in Fig. 3.6, environment affects the phenome in multiple ways, from its direct effects (e.g. toxins), through interactions with genes, to the modification of gene expression via epigenetic mechanisms. We should also keep in mind that environment has always been the key driver of natural selection, including *recent positive selection* of genetic variants (Text Box 3.6).

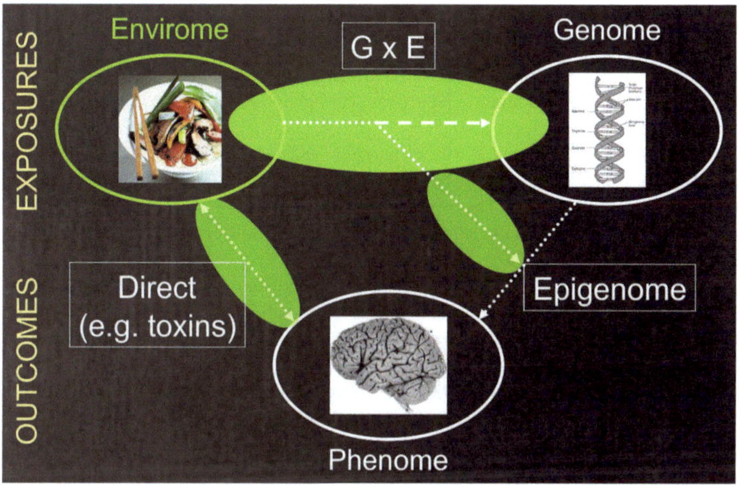

Fig. 3.6 Multiple ways in which environment affects the brain phenome. *G* genes; *E* environment

> **Text Box 3.6. Recent positive selection**
>
> The emergence of lactase persistence is but one example of recent positive selection. Mutations in the *MCM6* gene (located upstream of the lactase gene), underlying lactase persistence in adulthood, emerged in the context of dairy farming (and the availability of fresh milk) about 7,500 years ago (e.g. Itan et al. 2009). It is one of the clearest examples of recent environment-induced gene selection.

Assessment of the environment should never be dismissed, despite the difficulties (and possible inaccuracies) of its measurement and the inherent complexities of its interpretation, vis-à-vis phenome and genome. One would hope that new developments and advancements in "sampling" technologies, as well as an increased understanding of the epigenetic "signatures" of environmental exposures, might one day lead to the creation of envirome catalogues.

References

Abidin RR (1989) Parenting stress index manual. Pediatric Psychology Press, Charlottesville

Bernstein D, Fink L (1998) Childhood trauma questionnaire manual. The Psychological Corp, San Antonio

Bouchard TJ Jr, McGue M (2003) Genetic and environmental influences on human psychological differences. J Neurobiol 54(1):4–45

Cacioppo JT, Hawkley LC, Norman GJ, Berntson GG (2011) Social isolation. Ann N Y Acad Sci. doi:10.1111/j.1749-6632.2011.06028.x

Champagne FA, Curley JP (2009) Epigenetic mechanisms mediating the long-term effects of maternal care on development. Neurosci Biobehav Rev 33(4):593–600. doi:10.1016/j.neubiorev.2007.10.009

Cohen S, Mermelstein R, Kamarck T, Hoberman H (1985) Measuring the functional components of social support. In: Sarason IG, Sarason BR (eds) Social support: theory, research and application. Martinus Nijhoff, The Hague, Holland

Cohen S, Wills TA (1985) Stress, social support, and the buffering hypothesis. Psychol Bull 98(2):310–357

Freedman D, Thornton A, Camburn D, Alwin D, Young-demarco L (1988) The life history calendar: a technique for collecting retrospective data. Sociol Methodol 18:37–68

Gaser C, Schlaug G (2003) Brain structures differ between musicians and non-musicians. J Neurosci 23(27):9240–9245

Greenberg JR, Monson T, Gesino J (1993) University of Wisconsin family assessment Caregiver scale (UW-FACS): a new measure to assess families caring for a frail elderly member. J Gerontological Soc Work 19:49–68

Haghighi AMM, Bernard M, Abrahamowicz M, Leonard GT, Richer L, Perron M, Veillette S, Xu CJ, Greenwood CMT, Dias A, El-Sohemy A, Gaudet D, Paus T, Pausova Z (2012b) Opioid receptor mu 1 gene, fat intake and obesity in adolescence. Mol Psychiatry (in press)

Haghighi A, Schwartz DH, Abrahamowicz M, Leonard GT, Perron M, Richer L, Veillette S, Gaudet D, Paus T, Pausova Z (2013) Prenatal exposure to maternal cigarette smoking, amygdala volume, and fat intake in adolescence. JAMA Psychiat 70:98–105

Holt-Lunstad J, Smith TB, Layton JB (2010) Social relationships and mortality risk: a meta-analytic review. PLoS Med 7(7):e1000316. doi:10.1371/journal.pmed.1000316

House JS, Landis KR, Umberson D (1988) Social relationships and health. Science 241(4865):540–545

Itan Y, Powell A, Beaumont MA, Burger J, Thomas MG (2009) The origins of lactase persistence in Europe. PLoS Comput Biol 5(8):e1000491. doi:10.1371/journal.pcbi.1000491

James WP, Garza C (2012) Summary of the 24(th) Marabou symposium: nutrition and the human microbiome. Nutr Rev 70(Suppl 1):S87–S94. doi:10.1111/j.1753-4887.2012.00497.x

Kim DH, Petersen NM, Rahimi M, Burke J, Estrin D, Arab L (2008) Rewind: leveraging everyday mobile phones for targeted assisted recall. UCLA Technical Report

Kittleson MM, Meoni LA, Wang NY, Chu AY, Ford DE, Klag MJ (2006) Association of childhood socioeconomic status with subsequent coronary heart disease in physicians. Arch Intern Med 166(21):2356–2361. doi:10.1001/archinte.166.21.2356

Kuczynski J, Lauber CL, Walters WA, Parfrey LW, Clemente JC, Gevers D, Knight R (2011) Experimental and analytical tools for studying the human microbiome. Nat Rev Genet 13(1):47–58

Lerner RM, Lerner JV, Almerigi JB, Theokas C, Phelps E, Gestsdottir S, Naudeau S, Jelicic H, Alberts A, Ma L, Smith LM, Bobek DL, Richman-Raphael D, Simpson I, Christiansen ED, von Eye A (2005) Positive youth development, participation in community youth development programs, and community contributions of fifth-grade adolescents: findings from the first wave of the 4-H study of positive youth development. J Early Adolesc 25(1):17–71

Maguire EA, Gadian DG, Johnsrude IS, Good CD, Ashburner J, Frackowiak RS, Frith CD (2000) Navigation-related structural change in the hippocampi of taxi drivers. Proc Natl Acad Sci USA 97(8):4398–4403. doi:10.1073/pnas.070039597

Marmot MG, Syme SL (1976) Acculturation and coronary heart disease in Japanese-Americans. Am J Epidemiol 104(3):225–247

Marmot MG, Syme SL, Kagan A, Kato H, Cohen JB, Belsky J (1975) Epidemiologic studies of coronary heart disease and stroke in Japanese men living in Japan, Hawaii and California: prevalence of coronary and hypertensive heart disease and associated risk factors. Am J Epidemiol 102(6):514–525

Masten AS, Cicchetti D (2010) Developmental cascades. Dev Psychopathol 22(3):491–495. doi:10.1017/S0954579410000222

Northstone K, Joinson C, Emmett P, Ness A, Paus T (2011) Are dietary patterns in childhood associated with IQ at 8 years of age? A population-based cohort study. J Epidemiol Community Health. doi:jech.2010.111955 [pii] 10.1136/jech.2010.111955

Parker G (1990) The parental bonding instrument. Soc Psychiat Psychiat Epidemiol 281–282

Pea R, Mills M, Takeuchi L (2004) Making SENS: science education networks of sensors—report from an OMRON-sponsored workshop of the media-X program at Stanford University, October 3, 2003. Stanford Center for Innovations in Learning, Stanford

Plomin R (1994) The nature of nurture: the environment beyond the family. In: Plomin R (ed) Genetics and experience: the interplay between nature and nurture. Sage Publications Inc, Thousand Oaks, pp 82–101

Rogers I, Emmett P (1998) Diet during pregnancy in a population of pregnant women in South West England. ALSPAC study team. Avon longitudinal study of pregnancy and childhood. Eur J Clin Nutr 52(4):246–250

Scheibel A, Conrad T, Perdue S, Tomiyasu U, Wechsler A (1990) A quantitative study of dendrite complexity in selected areas of the human cerebral cortex. Brain Cogn 12(1):85–101

Schieman S, McMullen T (2008) Relational demography in the workplace and health: an analysis of gender and the subordinate-superordinate role-set. J Health Soc Behav 49(3):286–300

Sherbourne CD, Stewart AL (1991) The MOS social support survey. Soc Sci Med 32(6):705–714

Slimani N, Deharveng G, Unwin I, Southgate DA, Vignat J, Skeie G, Salvini S, Parpinel M, Moller A, Ireland J, Becker W, Farran A, Westenbrink S, Vasilopoulou E, Unwin J, Borgejordet A, Rohrmann S, Church S, Gnagnarella P, Casagrande C, van Bakel M, Niravong M, Boutron-Ruault MC, Stripp C, Tjonneland A, Trichopoulou A, Georga K, Nilsson S, Mattisson I, Ray J, Boeing H, Ocke M, Peeters PH, Jakszyn P, Amiano P, Engeset D, Lund E, de Magistris MS, Sacerdote C, Welch A, Bingham S, Subar AF, Riboli E (2007) The EPIC nutrient database project (ENDB): a first attempt to standardize nutrient databases across the 10 European countries participating in the EPIC study. Eur J Clin Nutr 61(9):1037–1056. doi:10.1038/sj.ejcn.1602679

Spanier GB (1989) Dyadic adjustment scale manual. Multi-Health Systems, North Tonawanda

Tremaroli V, Backhed F (2012) Functional interactions between the gut microbiota and host metabolism. Nature 489(7415):242–249. doi:10.1038/nature11552

Chapter 4
Genomics

This chapter provides an overview of concepts and tools relevant for studying the relationship between variations in the human genome and the brain and behaviour phenome. We will (1) present the basics of cellular and molecular genetics; (2) describe the main types of DNA variations; (3) explain the motivations and approaches for mapping genotype-phenotype associations; and (4) discuss issues related to the interpretation of such findings and their follow-up.

4.1 Chromosomes, Cell Division and Recombination

In humans, most of the genetic material is contained in DNA molecules packed into 23 different chromosomes. Most cells are **diploid**: They carry two sets of the 23 chromosomes (22 pairs of autosomal chromosomes and two sex chromosomes; $\sim$7 pg of DNA per cell). Sperm and egg cells are **haploid**: They contain only one set of 23 chromosomes.

Molecules of DNA are packed into chromosomes by forming complexes of DNA with histones (Text Box 4.1).

Text Box 4.1. Packaging of DNA: Euchromatin and heterochromatin

The "packaging" of DNA is achieved by coiling DNA molecules around the histone proteins (Fig. 4.1), resulting in a highly condensed *heterochromatin*. *Euchromatin* is a decondensed form of chromatin and, for the most part, is genetically "active": it can be transcribed to messenger RNA. In euchromatin, which is present in the nucleus throughout the majority of the cell cycle (interphase), the DNA–protein complex is characterized by weak bonds between DNA and H1 histone, as well as by a high level of the acetylation of the core histones. On the other hand, heterochromatin is genetically inactive. Constitutive (i.e. always condensed) heterochromatin is found in and around the centromeres and telomeres, while facultative (sometimes condensed, sometimes decondensed) heterochromatin provides,

T. Paus, *Population Neuroscience*, DOI: 10.1007/978-3-642-36450-1_4,

for example, one of the epigenetic mechanisms for regulating gene expression (see Chap. 5).

The tightness of these complexes is one of the key factors determining whether genes can be transcribed to RNA (turned on); the tighter the complexes are, the less likely the transcription of genes contained in the compacted segment of DNA (Fig. 4.2). During cell division, the entire chromosomes are tightly bundled up to facilitate their sorting into the daughter cells; no genes are transcribed in this phase of the cell cycle (Fig. 4.2, far right). The highly condensed form of chromatin is called heterochromatin, while the uncondensed form is euchromatin (Text Box 4.1). The uncoiled molecule of DNA present in euchromatin is ready for the synthesis of new DNA (replication) and RNA (transcription).

As the brain and body grow during development and old cells are replaced in adulthood, new somatic cells (all diploid) are generated through a typical form of **cell division**: **mitosis**. In the S phase of the cell cycle, DNA is replicated and the cell now contains a double amount of DNA ($\sim$14 pg); each of the 46 chromosomes now has its own homologue. These "old" and "new" chromosomes, tied together at a centromere, are referred to as "sister chromatids". In the M phase of the cycle, sister chromatids are separated—yielding a total of 92 chromosomes (i.e. four 23-chromosome sets) that are subsequently distributed into two daughter cells (46 chromosomes in each cell). It has been estimated that, over the course of

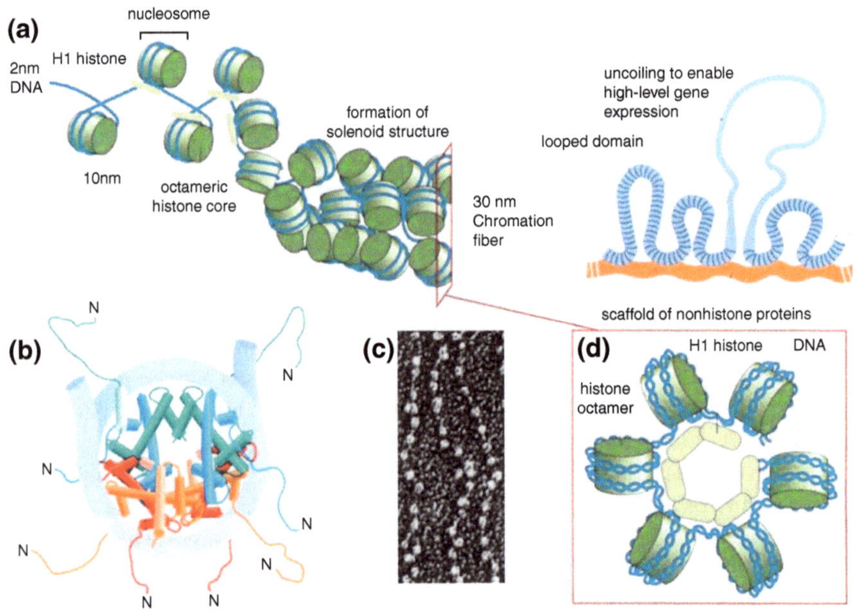

Fig. 4.1 DNA–chromatin complexes **a** binding of basic histone proteins, **b** a nucleosome—two turns of DNA wrapped around core histones, **c** electron micrograph of nucleosomal filaments, **d** cross section of a chromatin fibre. From Strachan and Read (2011)

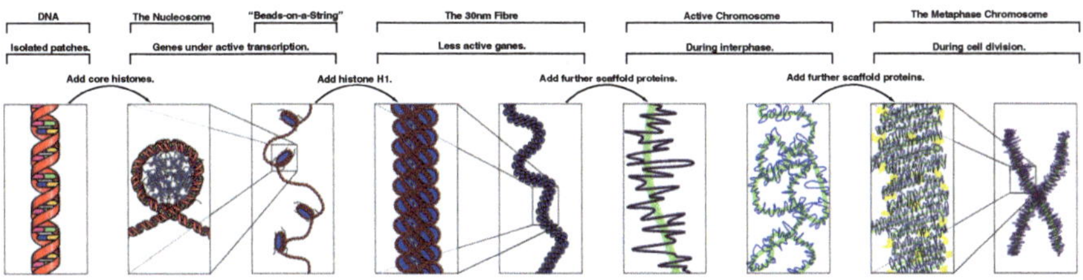

Fig. 4.2 Three levels of chromatin organization: (1) DNA wraps around histone proteins forming nucleosomes that look like beads on a DNA string; this is when genes can be transcribed (*turned on*); (2) multiple histones wrap into a 30 nm fibre consisting of nucleosome arrays in their most compact form; this is when genes in this segment of DNA are "turned off"; and (3) higher-level DNA packaging of the 30 nm fibre into the metaphase chromosome during cell division; this is when all genes are "turned off". From http://www.answers.com/topic/chromatin

the human lifespan, about 10^{17} mitotic cell divisions take place (Strachan and Read 2011, p. 32).

In the case of germ-line (diploid) spermatocytes and oocytes, new cells are generated through reduction division, **meiosis**, which produces haploid gametes (sperm and eggs; see Text Box 4.2).

Text Box 4.2. Production of haploid gametes: Meiosis

Meiosis progresses differently in the spermatocytes and the oocytes. For the spermatocytes, the first phase of meiosis (meiosis I) involves symmetric division of the (diploid) primary spermatocytes, resulting in two (diploid) secondary spermatocytes. In the second phase (meiosis II), another round of the symmetric cell division proceeds *without* DNA replication and generates *two haploid spermatids* per secondary spermatocyte. For the oocytes, meiosis I involves an *asymmetric division* that produces one diploid secondary oocyte and one polar body (which is discarded). In meiosis II, another asymmetric cell division, this time without DNA replication, produces one haploid mature egg (and one polar body, also discarded).

The meeting of an egg and a sperm represents a key event in the generation of genetic diversity through (1) an independent assortment of the maternal and paternal chromosomes, and (2) recombination.

Fertilization of a haploid egg by a haploid sperm results in a diploid zygote containing two sets of chromosomes—maternal and paternal. During meiosis I, the maternal and paternal homologues come together, forming the **bivalent**. After DNA replication, each of the homologous chromosomes of a given bivalent consists of two chromatids,[1] for a total of four strands of DNA. Next, the mitotic

[1] A chromatid is a copied chromosome that is paired with an identical chromosome at the centromere.

spindles pull one complete chromosome (i.e. two chromatids joined at centromere) towards each pole, which of the 23 chromosomes to go to which pole is random. Therefore, each daughter cell produced during the final phase of cell division inherits a *random* (independent) assortment of the maternal and paternal chromosomes (e.g. chromosomes 1, 4–6, 10, 12–14, 16, 17, 22, 23 and X from the mother, and chromosomes 2, 3, 7–9, 11, 15, 18–21 and X from the father). Given the number of chromosomes in one set (23), there are about 8,000,000 (2^{23}) possible combinations of the maternal and paternal chromosomes in the gametes (per meiosis; Strachan and Read 2011, p. 34).

The other key event occurring during meiosis is that of **recombination** (Fig. 4.3). As mentioned above, the duplicated maternal and paternal homologues form a bivalent consisting of four chromatids. Recombination (or crossover) occurs through a physical breakage at a corresponding location on two of the four strands, one maternal and one paternal, and the subsequent rejoining of the crossed-over fragments of the chromosomes. The recombined homologues remain connected at the point of a crossover, the **chiasma**, which is severed only when the mitotic spindles start pulling the chromosomes towards the poles. There are ~55 chiasmata per cell during the male meiosis (but 50 % more during the female meiosis), thus indicating the frequency of recombination during sexual reproduction (Strachan and Read 2011, p. 35).

Together, the two mechanisms of combining maternal and paternal genetic material represent a major source of genetic variability in the human population; one ensuring a combination of genes located on different chromosomes (**chromosomal assortment**) and the other mixing maternal and paternal genes within a chromosome (**recombination**).

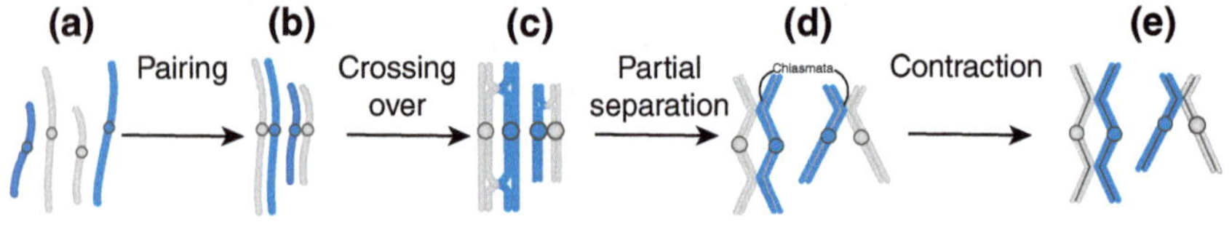

Fig. 4.3 Meiosis: an example of two chromosomes from the mother (*light blue*) and the father (*dark blue*) **a** Chromosomes are duplicated (chromatids) but remain unpaired, **b** duplicated homologues of the maternal and paternal chromosome (two chromatids each) pair up to form a bivalent (four chromatids), **c** recombination (crossing over): physical breakage and subsequent rejoining of maternal and paternal chromosome fragment (in this example, there are two crossovers in the bivalent on the *left side* and one in the bivalent on the *right*), **d** homologous chromosomes separate slightly, except at the chiasmata (points of crossover), and **e** bivalents contract and transit to metaphase. From Strachan and Read (2011)

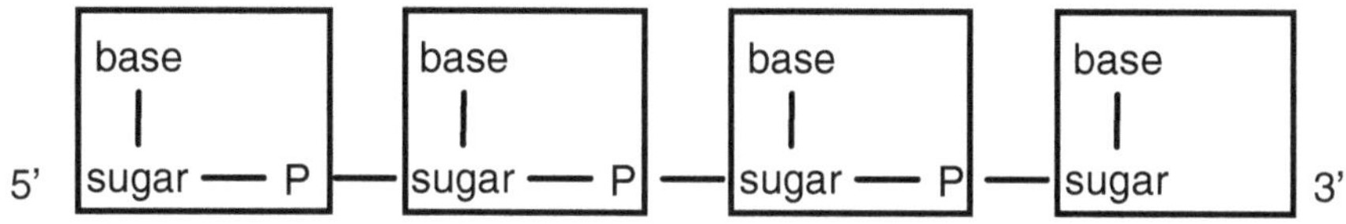

Fig. 4.4 Two building blocks of a DNA molecule: a base (A, C, G or T) and a sugar (deoxyribose) linked to the next sugar by a phosphate (*P*). Each nucleotide (indicated by a rectangle) consists of a base, a sugar and a phosphate. The 5′ (five prime) and 3′ (three prime) ends refer, respectively, to the 5th and 3rd carbon in the sugar. Modified from Strachan and Read (2011)

4.2 Genetic Code, Gene Transcription and Translation

Let us now focus on the actual DNA molecule and talk about the **genetic code**.

The DNA molecule is a double helix in which two (complementary) strands of DNA are bound to each other through nitrogenous bases (**base pairs**). In DNA, the bases are adenine (A), cytosine (C), guanine (G) and thymine (T); the canonical (Watson–Crick) base pairings are A–T and G–C (Watson and Crick 1953). The backbone of each strand is made of a five-carbon sugar (deoxyribose) linked to the next sugar with a phosphate (see Figs. 4.4 and 1.2).

Nucleotides are the basic DNA units; each nucleotide consists of a base, a sugar and a phosphate. The genetic code consists of nucleotide triplets (e.g. ATG) that are transcribed to messenger RNA (mRNA) as three-letter **codons**. In mRNA, each of 64 possible codons corresponds to one of 20 amino acids and to one of three so-called STOP codons. One codon (AUG) codes an amino acid (methionine) and also indicates where translation into a protein begins (START codon). Table 4.1 contains a DNA codon table, indicating DNA bases on the "sense" DNA strand (see below).

Clearly, the genetic code must be read in the correct direction. During the synthesis of both the new DNA (replication) and RNA (transcription), the DNA and RNA polymerases copy the code of the template strand of DNA in the $5' \rightarrow 3'$ direction; the 5′ and 3′ ends have sugar residues in which carbons number 5 and 3, respectively, are not linked to another sugar. The nucleotide sequence of RNA transcript (mRNA) is complementary to the template (sense) strand of DNA and, as such, it is identical to the sequence of the non-template (anti-sense) strand (with one exception: thymine is replaced by uracil).

All cells in the body contain the same DNA. Whether a particular cell synthesizes proteins that turn it, for example, into a pyramidal neuron or a Chandelier interneuron depends mainly on which of its genes are transcribed into messenger RNA. Gene *transcription* is regulated by **transcription factors**, a family of proteins that bind to a particular DNA sequence elements—a **promotor**—located

Table 4.1 DNA codon table

Genetic code									
1st base	2nd base								3rd base
	T		C		A		G		
T	TTT	(Phe/F) Phenylalanine	TCT	(Ser/S) Serine	TAT	(Tyr/Y) Tyrosine	TGT	(Cys/C) Cysteine	**T**
	TTC		TCC		TAC		TGC		**C**
	TTA	(Leu/L) Leucine	TCA		TAA	Stop	TGA	Stop	**A**
	TTG		TCG		TAG	Stop	TGG	(Trp/W) Tryptophan	**G**
C	CTT		CCT	(Pro/P) Proline	CAT	(His/H) Histidine	CGT	(Arg/R) Arginine	**T**
	CTC		CCC		CAC		CGC		**C**
	CTA		CCA		CAA	(Gln/Q) Glutamine	CGA		**A**
	CTG		CCG		CAG		CGG		**G**
A	ATT	(Ile/I) Isoleucine	ACT	(Thr/T) Threonine	AAT	(Asn/N) Asparagine	AGT	(Ser/S) Serine	**T**
	ATC		ACC		AAC		AGC		**C**
	ATA		ACA		AAA	(Lys/K) Lysine	AGA	(Arg/R) Arginine	**A**
	ATG[A]	(Met/M) Methionine	ACG		AAG		AGG		**G**
G	GTT	(Val/V) Valine	GCT	(Ala/A) Alanine	GAT	(Asp/D) Aspartic acid	GGT	(Gly/G) Glycine	**T**
	GTC		GCC		GAC		GGC		**C**
	GTA		GCA		GAA	(Glu/E) Glutamic acid	GGA		**A**
	GTG		GCG		GAG		GGG		**G**

The DNA codons occur on the sense DNA strand and are arranged in a $5' \rightarrow 3'$ direction upstream[2] from a gene and in its immediate vicinity. Once bound to a promotor,[3] a transcription factor guides the RNA polymerases transcribing the template strand of DNA into RNA. In addition to promotors, transcriptional activity can be enhanced or inhibited by "enhancers" or "silencers", respectively, as well as by a number of epigenetic mechanisms (see Chap. 5). Once a full RNA transcript is synthesized, a set of steps ensues that it makes a "mature mRNA". One of the key steps in this chain of reactions is **RNA splicing**, whereby the non-coding parts of the gene (**introns**) are removed and the remaining coding sequences (**exons**) are tied together to form a shorter mRNA (Text Box 4.3).

Text Box 4.3. Exons and introns

An exon (a DNA region that will be *ex*pressed) is part of DNA that is transcribed to mRNA and, in most cases, translated into a protein. An intron (intragenic region) refers to a DNA sequence within a gene that is removed (by splicing) during transcription and, therefore, is not the part of the final mRNA. Typically, an intron is recognized by the fact that it starts with a GT and ends with an AG (the GT–AG rule). For the majority of multi-exon human genes (Pan et al. 2008), this process may bring together a slightly different subset of exons. Such an "alternative splicing" may result, after translation, in different forms of the same protein.

The final step on the road from DNA to protein is *translation.* Unlike the preceding steps, which all take place in the cell nucleus, translation occurs on ribosomes, located in the cell cytoplasm (Fig. 1.2). This is where the genetic code is translated into a polypeptide: an RNA codon into an amino acid. As we see in Table 4.1, most amino acids are coded by more than one codon, hence, "degeneracy" of the genetic code. Translation results in a chain of amino acids—a polypeptide—that is often modified (during or after translation) by other chemical processes, such as phosphorylation, methylation or acetylation. The final product—a protein—consists of one or more polypeptides, which may undergo further post-translational modifications, co-determining the ultimate structure (and hence, functionality) of the protein.

Overall, given the above rules governing replication and transcription of DNA, it is not surprising that one letter of the genetic code can make a big difference in the final outcome: the amount and structure (and therefore function) of a protein.

[2] Relative positions of structures along a strand of nuclei acid are typically referenced to the 5′ (upstream) and 3′ (downstream) ends.

[3] Most promotors consist of short DNA sequences, such as TATA box (e.g. TATAAA), GC box (usually GGGGCGG) or CAAT box (Strachan and Read 2011).

4.3 DNA Variations

Variations in DNA inherited through the germ line from our parents represent the main molecular mechanism underlying the heritable portion of inter-individual variability in brain and behaviour. In general, one can distinguish between (1) *single-nucleotide variations*, which involve base substitutions, deletions and insertions of a single nucleotide; and (2) *multiple-nucleotide variations*, which include insertions and deletions (so-called **indels**) of multiple nucleotides, as well as copy-number variations (CNVs) and inversions. In current genetic mapping studies (see Sect. 4.4), the most commonly employed DNA variations are **single-nucleotide polymorphisms (SNPs)** and **CNVs**.

SNPs are variations in the nitrogenous base of a single nucleotide (e.g. A instead of G) located anywhere in the DNA sequence. It has been estimated that any two individuals would differ, on average, in one out of 1,200 bases; with three billion bases per 23 chromosomes, this represents about 2,500,000 SNPs distinguishing any two individuals (http://hapmap.ncbi.nlm.nih.gov). About 10 million "common" DNA variants (primarily SNPs) occur with a frequency higher than 1 % across various human populations, as sampled by the Human Genome Project, the SNP Consortium and the International HapMap Project (The International HapMap 3 Consortium 2010). The latest version of the SNP database (dbSNP Build 135, human Genome Build 37.3; http://www.ncbi.nlm.nih.gov/SNP/) contains a total of 41,740,143 validated SNPs (reference SNP [rs] ID numbers), of which about 22 million are located within and/or near genes[4] and about 19 million are found in intergenic regions.

When located in exons ($\sim$3 million out of all 41 million SNPs), a SNP can be either non-synonymous or synonymous; the former refers to the fact that the sequence substitution results in a different amino acid, whereas this is *not* the case for the latter (recall the "degeneracy" of the genetic code, see Table 4.1). For example, the change of "G" to "A" in the first letter of a DNA codon for valine (GTG) results in a codon that codes instead for methionine (ATG; see Table 4.1). This kind of genetic variation is found, for example, in a commonly studied SNP (rs6265) of the *BDNF* gene (Text Box 4.4).

Text Box 4.4. A non-synonymous SNP in the *BDNF* gene

A commonly studied SNP (rs6265) in *BDNF*—the "G196A" DNA polymorphism—results in a "val66met" (valine to methionine) polymorphism in the proBDNF polypeptide. The number between G and A refers to the position of the nucleotide in cDNA, and the number between val and met refers to the position of the amino acid in the polypeptide/protein. This polymorphism appears to affect intracellular packaging of proBDNF, its

[4] Note that only $\sim$3 million SNPs are located in exons, the rest are in introns.

axonal transport and, in turn, activity-dependent secretion of BDNF at the synapse (Chen et al. 2004).

Although it is more likely for the non-synonymous SNPs to influence the ultimate function of a given gene product, this is also possible for the synonymous SNPs and SNPs located in introns or intergenic regions. Even though synonymous SNPs do not change amino acids, they may affect the function by influencing, say, gene expression. The distinction between "functional polymorphisms" and "markers" will be discussed in the next Sect. 4.3.

CNVs refer to various *quantitative* variations in the genome, including tandem repeats, deletions and duplications; they can vary in size between ~1 kb and 1 Mb (Text Box 4.5).

Text Box 4.5. Variable Number of Tandem Repeats (VNTR)

VNTR is an example of a CNV commonly used in genetic studies. Historically, VNTRs were used in the first genome-wide studies of complex traits. Some of VNTRs are functional polymorphisms; for example, we use the variable number of CAG triplets in Exon 1 of the androgen receptor gene in this context (see Fig. 9.4).

In a survey of CNVs larger than 500 bp, about 8,000 different CNVs were revealed in DNA from 40 individuals. As shown in Fig. 4.5, the majority of these genetic variations were found in intergenic regions (Conrad et al. 2010), where they might influence expression of the genes located in their vicinity. For example, many developmental genes are flanked by large intergenic regions (gene deserts) containing enhancers of gene expression (Klopocki and Mundlos 2011). CNVs might be behind some of the associations previously observed between SNPs and complex traits: almost a third of SNPs associated with a complex trait are in LD with a CNV (Conrad et al. 2010). This opens up the possibility that such SNPs mark the locations of "causal" CNVs.

To appreciate how a person's genome might differ from the "average" genome, we can compare the full DNA sequence of an individual with a reference assembly.[5] This has been done for the founder of Celera (https://www.celera.com), Craig Venter's own DNA sequence. Venter's genome differs from the reference assembly in the following features: 3.2 million SNPs; 292,000 heterozygous insertion/deletion variants (indels); 559,000 homozygous indels; 90 large inversions; and 62 large copy-number variants. Almost 44 % of Venter's genes had a sequence variant, with 17 % of them encoding an altered protein (Levy et al. 2007). This level of knowledge of the human genome represents an extraordinary platform from which to embark on mapping genotype-phenotype associations.

[5] The National Center for Biotechnology Information human reference assembly.

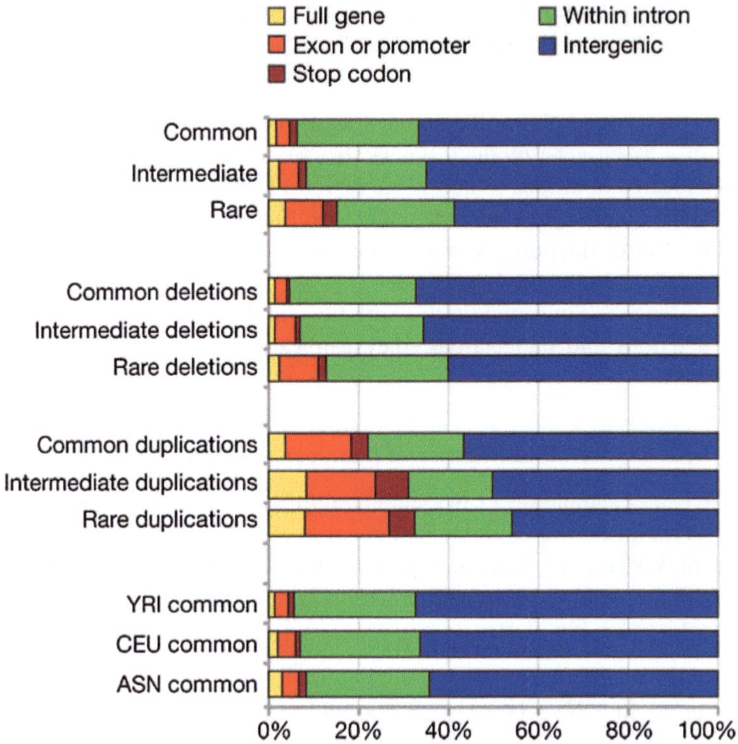

Fig. 4.5 Copy-number variations. Population frequency classes: common (MAF > = 0.1), intermediate (0.1 > MAF > 0.01) and rare (MAF < = 0.01). YRI, Yoruba from Ibadan, Nigeria; CEU, Utah residents with ancestry from northern to western Europe; ASN, denotes JPT (individuals in Tokyo, Japan) and CHB (individuals in Beijing, China). From Conrad et al. (2010)

4.4 Mapping Genotype-Phenotype Associations: Hypothesis-Driven Approach

In principle, there are two reasons for using genomic knowledge in population neuroscience: to (1) gain insight into the molecular mechanisms (pathways) underlying a particular systems-level phenomenon using genetic variations with known function and (2) discover novel associations between genes and brain/behaviour phenotypes.

A structural and/or functional neuroimaging study is but a starting point that leads to new knowledge. For example, an MR-based morphometric study might reveal a group difference in the volume of the hippocampus between patients with post-traumatic stress disorder (PTSD) and healthy controls. There is a clear need to follow up such initial (descriptive) findings by asking a series of mechanistic questions. For example, is the group difference due to the activation of the hypothalamus–pituitary–adrenal (HPA) axis and the related central release of cortisol? If yes, which one of the variety of possible downstream effects of cortisol—such as decreased neurogenesis, increased cell death or impoverished dendritic branching (Conrad 2010; Zheng et al. 2006)—plays the most significant role?

Unlike studies carried out in experimental animals, human studies are limited to the use of non-invasive techniques; we do not have direct access to the brain tissue of living individuals (for detailed microscopic or chemical analyses) and cannot manipulate different molecular systems before, during and after stress exposure. This is where inter-individual variability in genes relevant for different molecular processes may come into play. In this context, the most common strategy is to select a candidate gene physiologically relevant for the question at hand—say, the glucocorticoid receptor gene (*NR3C1*)—and then identify its variant associated with a known difference in the function of this gene, the **functional polymorphism.** This allows us to explore a moderating effect of this polymorphism on the observed exposure–outcome relationship and, in this way, test the involvement of a particular molecular pathway in the relationship (see Sect. 9.2 for discussion on Mendelian randomization as a tool for injecting causality into observational studies).

A functional polymorphism refers to a genetic variant that causes differences in gene expression or in the structure/function of the final gene product (e.g. alternative splicing, resulting in a less functional form of the protein). In the case of *NR3C1*, an amino acid substitution of arginine for lysine (codon 32 of exon 2) results in a shift to a less active form of the glucocorticoid receptor, namely the GR-A variant (Russcher et al. 2005). We can now use this (*NR3C1*) functional polymorphism and test whether or not stress—via stress-induced cortisol effects (exposure)—mediated by the glucocorticoid receptor is indeed associated with the hippocampal volume (outcome). If this is the case, PTSD patients with the less

Fig. 4.6 Searching the AmiGO gene ontology (*GO*) database for gene products associated with "neurogenesis"

active GR-A variant should have hippocampal volumes comparable with those of healthy controls.

How do we go about putting together a list of relevant candidate genes? First of all, we need to identify biological processes that are likely to be involved in the formation of our phenotype. This step can start in a narrow way (e.g. HPA-related enzymes and receptors) and expanded later to include other related processes (e.g. neurogenesis, apoptosis, dendritic growth). Following a background reading about the pertinent cellular pathways, one can take advantage of a number of bioinformatics tools to find the relevant "families" of genes. For example, the AmiGO gene ontology (GO) database (http://amigo.geneontology.org) allows one to identify gene products associated with a given biological process. Thus, entering "neurogenesis" as a GO term yields four results, one of these (regulation of neurogenesis) includes 577 gene products in humans (see Fig. 4.6).

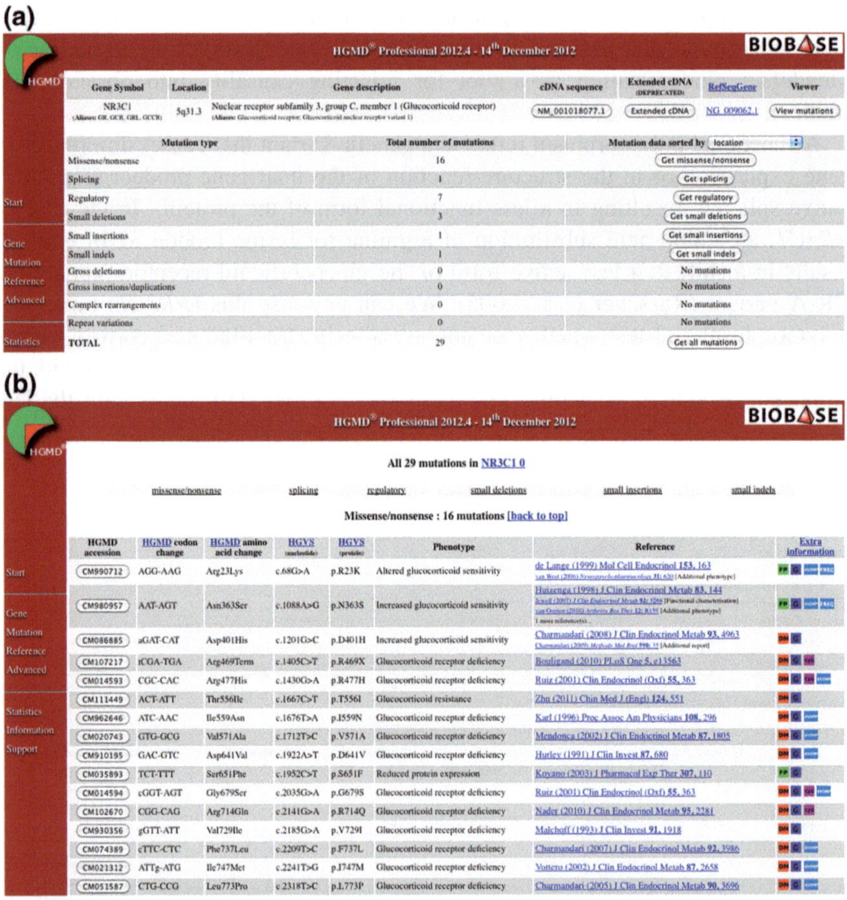

Fig. 4.7 **a** and **b**: Human Gene Mutation Database. Searching for mutations in the glucocorticoid receptor gene (*NR3C1*)

The next step involves identifying functional polymorphisms for each of the genes of interest. Again, there are bioinformatics tools that can be used for this purpose. For example, the **Human Gene Mutation Database** contains about 120,000 entries of disease-associated and/or functional polymorphisms (mutations) reported in the literature (Stenson et al. 2009; http://www.hgmd.org/); the majority of these polymorphisms are non-synonymous sequence substitutions (**missense mutations**) and small deletions. Entering "*NR3C1*" in this database yields 29 entries (Fig. 4.7a), with relevant details for each entry (Fig. 4.7b). In the absence of a known functional polymorphism, one can identify non-synonymous SNPs and select only those that are predicted to affect protein function; this can be done, for example, with the **Sorting Tolerant From Intolerant** (**SIFT**) algorithm (Kumar et al. 2009; http://sift.jcvi.org/).

In summary, this "functional polymorphism" approach is helpful for testing initial ideas about the possible molecular pathways underlying the phenomenon of interest, thus providing mechanistic insights otherwise unavailable in the living human. Naturally, the sample size will dictate how many functional polymorphisms can be tested in a given study. This is why selecting only *functional* polymorphisms—rather than any DNA variations—might be a sensible starting point.

4.5 Mapping Genotype-Phenotype Associations: Hypothesis-Free Approach

By definition, a "hypothesis testing" approach described in the preceding section works only when we have a workable hypothesis about some of the biological processes underlying a particular systems-level phenotype. Although this can be a satisfying approach vis-à-vis the traditional scientific method of making observations and doing experiments, we also need to make new observations that are unconstrained—at least initially—by prior knowledge. Searching for new genotype-phenotype associations can be viewed as such a "hypothesis-generating" enterprise. Given that the majority of brain and behaviour phenotypes show a reasonable level of heritability (e.g. Bouchard and McGue 2003; Giedd et al. 2007; Peper et al. 2007), this search is guided by an overarching hypothesis, namely that a genetic variation underpins the heritable portion of inter-individual variability in these complex traits. Looking for DNA variations associated with such traits represents one possible strategy in these explorations.

The first goal of this exploratory search for new associations between the genes and brain/behaviour is to find, on the genome, locations associated with the phenotype of interest. In this context, genetic variants are used simply as *markers* on the physical map of the genome, telling us that genetic variations in the vicinity of a particular DNA locus are linked (or associated) with the variations in the phenotype. There are two general approaches to identify such genotype-phenotype relationships: **linkage** and **association studies**.

Table 4.2 Criteria for reporting linkage in genome-wide studies

Category of linkage	Expected number of occurrences by chance in a genome-wide scan	Range of approximate *p* values	Range of approximate LOD scores
Suggestive	1	7×10^{-4}–3×10^{-5}	2.2–3.5
Significant	0.05	2×10^{-5}–4×10^{-7}	3.6–5.3
Highly significant	0.001	$<=3 \times 10^{-7}$	>=5.4
Confirmed[a]	0.01		

[a] In a search of a candidate region that gave significant linkage in a previous independent study. From Strachan and Read (2011)

Linkage-based studies can be conducted only in families; here, we trace the simultaneous co-segregation of the genetic variability (at a given location) and the phenotype across generations of related individuals (and/or between siblings). This approach has been extremely successful in the search for genes associated with monogenic (Mendelian) disorders, such as Huntington's disease (Gusella et al. 1983). It has been also used—albeit with less success—in the initial genetic studies of polygenic (non-Mendelian/complex) disorders, such as schizophrenia (Altmüller et al. 2001). In the 1990s, the Human Genome Project expanded the tools for linkage studies in the form of microsatellites, the most common of which are so-called short tandem repeats, consisting of 10 or more repetitions of two to six nucleotides (e.g. CAG), located along the genome and positioned on the genome's physical map using the unique sequence-tagged sites (see Sect. 2.2). The microsatellites tend to be highly polymorphic in the number of repeats. In linkage studies, this polymorphism allows one to track the co-segregation—across generations or between siblings—of the variability in a trait with the variability in a particular microsatellite. In this manner, one identifies on the genome a physical location associated with a particular pattern of trait variations in a set of families or sib pairs. Results of linkage studies are typically reported using LOD (logarithm of odds) scores.[6] Table 4.2 contains criteria suggested by Lander and Kruglyak (1995) for reporting linkage in genome-wide studies (from Strachan and Read 2011): a highly significant linkage—expected to occur only 0.001 times by chance in a genome-wide study—should have a LOD score ≥ 5.4 (and $p \leq 3 \times 10^{-7}$).

Typical linkage studies scan genomes in rather large "chunks" (~5–10 Mb segments), reflecting the density of microsatellites but also, more importantly, the fact that family-based linkage works with rather large segments of the genome shared by relatives. This is because a limited number of recombinations occur during meiosis in the parental germ-line cells. Genome-wide association studies (GWAS) take a very different approach.

[6] A LOD score of 3 indicates that the probability of finding a linkage is 10^3 higher than the null hypothesis (i.e. no linkage). High LOD scores are less likely and, as such, are associated with low *p* values, indicating the low probability of observing them by chance.

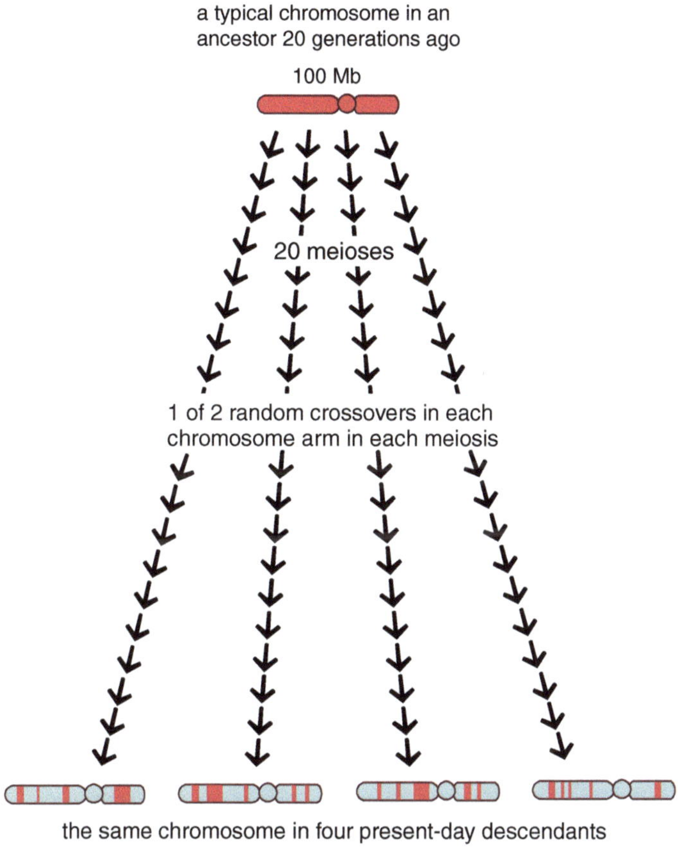

Fig. 4.8 The size of shared ancestral segments of a chromosome. For each meiosis (one meiosis per generation), there will be about one or two random crossovers in each arm of the chromosomes. Therefore, only a small proportion of the DNA sequence of the ancestral chromosome will be inherited by descendants after 20 generations (*red* segments). And even smaller proportion of all descendents will share the ancestral segments. From Strachan and Read (2011)

In *association-based studies*, we simply search for statistical associations between a specific allele, usually a SNP, and a phenotype. This search tends to be carried out in a population of unrelated individuals (but can also be conducted in families). The main assumption here is that a specific (causal) genetic variation—which emerged through hundreds or thousands of meioses across multiple generations in a given population—was preserved through the process of natural selection and, as such, confers a certain survival value. In theory, one can search for a genetic variation—a single-nucleotide substitution in the case of SNPs—by testing statistically for associations between the allele (e.g. A, G) and the phenotype at each of the three billion bases constituting the human genome. But, as

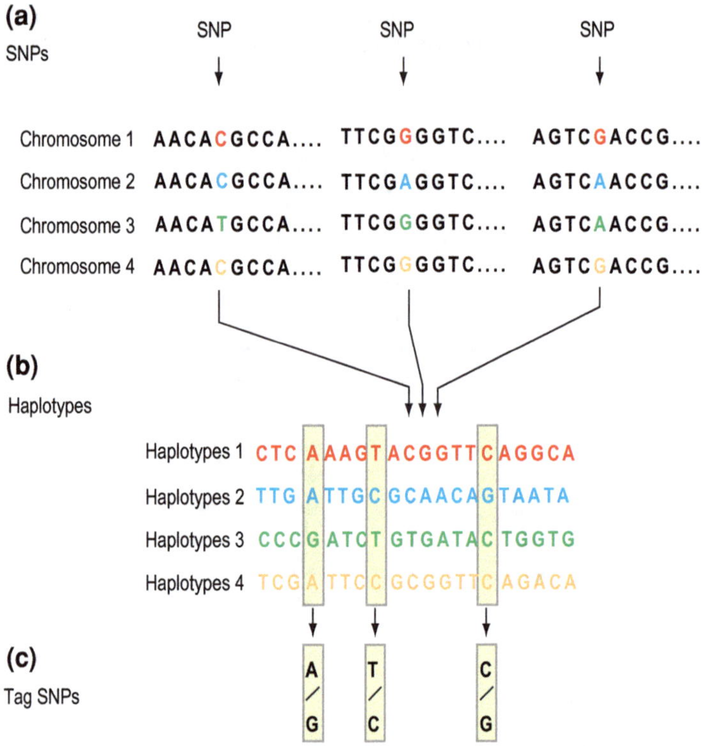

Fig. 4.9 Constructing the haplotypes. "The construction of the HapMap occurs in three steps: **a** SNPs are linked in DNA samples from multiple individuals, **b** adjacent SNPs that are inherited together are assembled into haplotypes, **c** tag SNPs within haplotypes are identified, uniquely characterizing those haplotypes. By genotyping the three tag SNPs shown in this figure, researchers can identify which of the four haplotypes shown here are present in each individual." From: hapmap.ncbi.nlm.nih.gov/whatishapmap.html.en

we have learned above (Sect. 4.3), this is not necessary: "only" about 10 million SNPs occur in the human genome with a frequency higher than 1 %. Furthermore, individual SNPs are not independent of each other: they come in blocks. These blocks reflect the recombination history of the population. Thus, in a span of 500 years, a "uniform" ancestral chromosome at Generation 1 becomes a "striped" chromosome in Generation 22 (Fig. 4.8).

Using data of the HapMap project (see below), it has been estimated that the average size of such blocks is about 10 kb; individuals who come from genetically older populations (e.g. the Yoruba people from Nigeria) have shorter blocks (7.3 kb/block) than those coming from a more recent population (e.g. individuals of European ancestry from Utah; 16.4 kb/block). The International HapMap Project (http://hapmap.ncbi.nlm.nih.gov/; The International HapMap 3 Consortium

2010) has identified block structures in four population samples,[7] thus identifying haplotypes (Text Box 4.6) in these populations.

Text Box 4.6. Haplotypes

Haplotypes refer to a combination of alleles that are inherited together (Fig. 4.9). Note that haplotypes do not simply contain alleles that are physically close to each other. Due to the varied rate of recombinations at different chromosomal locations (e.g. high in the recombination "hot spots" and very low near centromeres), adjacent alleles may belong to different haplotypes and physically distant alleles can belong to the same haplotype.

Once the haplotypes are identified, one can proceed with selecting so-called tag SNPs. These tag SNPs are in strong linkage disequilibrium (LD) with other SNPs located in a block of DNA and they are inherited together (see Fig. 4.9). Given the "block" structure of the genome, the number of such tag SNPs is much lower (between 300,000 and 600,000) than the number of common SNPs (10 million).

In the 1998 HGP report, Collins et al. (1998) predicted the identification of 100,000 SNPs in the genome over the next 5 years. The subsequent development of DNA microarrays (or DNA "chips") by companies such as Affymetrix and Illummina has increased the number of SNPs per DNA sample while reducing the price dramatically. Today, one can obtain genome-wide coverage with 800,000 to five million SNPs, for $200–$600 per sample. Positioning of SNPs on these chips is informed by advances made by projects such as the HapMap, thus adjusting the coverage to capture the greatest amount of genetic variation.[8]

A typical protocol used to obtain genome-wide coverage with this approach begins with taking a blood sample, spinning it to separate plasma and blood cells, and using the latter for DNA extraction. Typically, one needs from 200 to 500 ng of DNA.[9] With DNA extracted, the first step involves whole-genome amplification, fragmentation and hybridization of the DNA fragments to 50-mer[10]

---

[7] Four samples studied by the International HapMap Project: 30 Utah parent–child trios, originally from northern and western Europe (CEU), 30 Yoruba parent–child trios from Ibadan, Nigeria (YRI), 45 Han Chinese individuals from Beijing (CHB) and 45 Japanese individuals from Tokyo (JPT).

[8] For example, Illumina HumanOmniExpress BeadChip contains 730,225 SNPs. Of these, 392,197 are SNPs located within 10 kb of the RefSeq genes and 15,062 are non-synonymous (NCBI annotated) SNPs.

[9] Given that one diploid cell contains ~7 pg of DNA, we need ~30,000 cells to get 200 ng of DNA; 1 ml of blood should contain $\sim 4 \times 10^6$ white cells, of which $\sim 10^6$ are mononuclear leukocytes, that is, lymphocytes and monocytes.

[10] In general, "mer" refers to a repeat unit (poly*mer*). In this case, it refers to the number of oligonucleotides (oligo*mers*).

complementary probes on the chip, and labelling all bonded (hybridized) nucleotide sequences with fluorescent "target" sequences. The strength and colour of the fluorescent signal at a given "spot" indicate which of the nucleotides are present at a given physical location in the sampled genome. The next step involves quality control and, in turn, selection of the final set of valid SNPs for statistical analysis (Anderson et al. 2010). This step identifies (and excludes) all SNPs with poor "call rate"[11] and all individuals with more than 3 % of missing SNPs (a genotype failure rate).[12] Then, one excludes SNPs that depart (at a significance level of p between 10^{-4} and 10^{-6}) from the Hardy-Weinberg equilibrium, assuming that allele and genotype frequencies in a population remain constant.[13] Finally, one calculates the so-called minor allele frequency (MAF) and, typically, excludes SNPs with MAF lower than 1–2 % (i.e. rare variants).[14]

Now that we have a final set of SNPs, we can test for associations between the genotype and phenotype across the genome. This analysis can be carried out with a number of tools, such as PLINK (http://pngu.mgh.harvard.edu/~purcell/plink/) or KING (http://people.virginia.edu/~wc9c/KING/). Results, in the form of p values, indicating statistical significance of the tested associations, can be visualized using tools such as WGAViewer (http://compute1.lsrc.duke.edu/softwares/WGAViewer/). Given the very high number of tests carried out in genome-wide scans (~800,000 for a DNA chip yielding 800,000 SNPs), the cut-off threshold for a significant finding is very high (7.2×10^{-8} suggested by Dudbridge and Gusnanto 2008; but see Pe'er et al. 2008 for alternatives). Once a significant association is found, the next phase begins.

4.6 Follow-up of GWAS Findings

Given the growth of GWAS of various phenotypes, any finding should be first checked against those deposited in the catalogue of published GWA (Hindorff et al.; http://www.genome.gov/gwastudies/); as can be seen in Fig. 4.10, the number of GWAS reports has risen exponentially over the past 5 years. This catalogue can be searched in a variety of ways, including gene names, SNPs or diseases/traits.

In the case of novel (previously unreported) associations, a replication (and meta-analysis, that is, calculating a p value across the discovery and replication

---

[11] Call rate indicates the proportion of genotypes (per SNP) with non-missing data; call rate >= 95 % is typically acceptable.

[12] This is usually due to the poor quality of a DNA sample.

[13] At a bi-allelic locus in Hardy–Weinberg equilibrium, these probabilities are as follows: $(1-q)^2$ for aa, $2q(1-q)$ for aA and g^2 for AA (q is the minor allele probability).

[14] By definition, less common (MAF < 1 %) variants will be present only in a few individuals, thus not afford the necessary statistical power for detecting genotype-phenotype associations with a modest effect size.

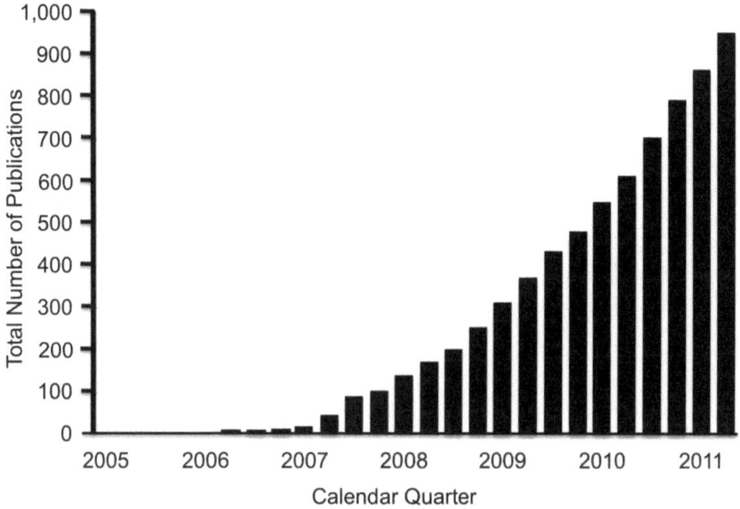

Fig. 4.10 Number of published reports of genome-wide association studies (from http://www.genome.gov/gwastudies/index.cfm?pageid=26525384#searchForm)

analyses) should be sought when possible; a successful replication requires finding an effect with nominal evidence against the null hypothesis ($p < 0.05$) and having the direction consistent with that observed in the discovery sample (NCI-NHGRI Working Group on Replication in Association Studies 2007; see also Colhoun et al. 2003 and Igl et al. 2009 for further discussion on this topic). In the next step, we should ask about the type of identified SNPs with regard to their possible consequences on the function of the implicated gene. Are they located in the promotor or the coding region? If the latter, are they synonymous or non-synonymous? Are they in LD with known functional variants? These questions are the same as those addressed when discussing functional polymorphisms in the beginning of Sect. 4.4. If the function of the associated gene is unknown, or we need further insights about the possible molecular pathways underlying the association between the gene and the phenotype, it may be useful to explore so-called co-expression networks, using available databases of gene expression and tools such as ErmineJ (http://www.chibi.ubc.ca/ermineJ/; Lee et al. 2005; Gillis et al. 2010; Gilles and Pavlidis 2011).

Any genotype-phenotype association represents only a statistical correlation between two domains and, as such, lacks *causality*. The latter can be determined only by manipulating the genome of experimental animals and/or tissue cultures. A variety of approaches are available in this context, such as knock-in and knock-out versions of a given gene or the silencing of gene expression during a particular developmental period using an RNA interference.

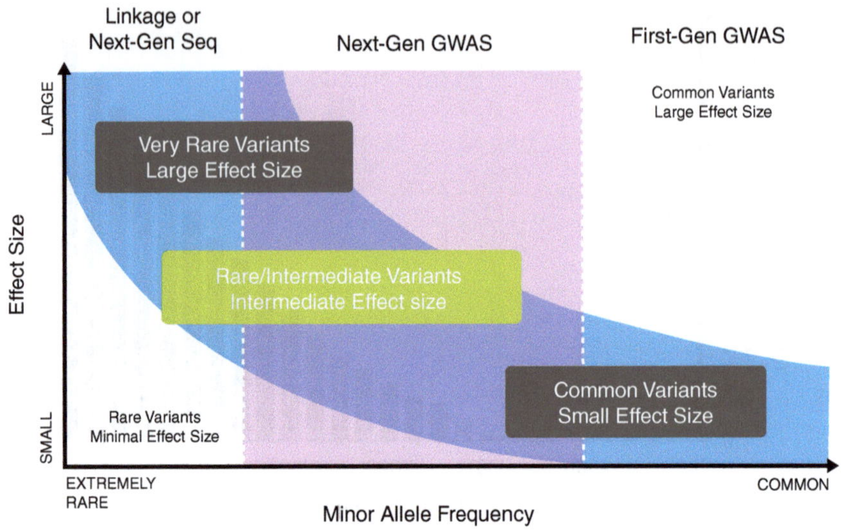

Fig. 4.11 A theoretical relationship between minor allele frequency and effect size

4.7 Missing Heritability

Twin studies suggest fairly high values of heritability (>50 %) for many brain (e.g. Glahn et al. 2007) and behaviour (e.g. Bouchard and McGue 2003) phenotypes. And yet, if we were to add all genetic variations (as revealed by genome-wide studies) of some of these complex traits, the sum of the variance explained by these SNPs would not reach these twin-based heritability values—hence the enigma of missing heritability. A number of possible explanations have been put forward.

First of all, GWAS are based on the "common trait–common variant" paradigm. But as can be seen in Fig. 4.11 (bottom right), the common variants usually explain only a small proportion of variance in a given complex trait. In the case of height—a highly (~80 %) heritable trait and one that has been a target of multiple GWA studies—well-replicated SNPs, combined together, explain only ~5 % of variance (Visscher 2008). And yet, as shown in the elegant studies of Visscher and colleagues, if one considers all genome-wide SNP variations (as a proxy of "relatedness"), then genetic similarity across unrelated individuals (captured by ~500,000 SNPs) explains ~45 % of variability in their height (Yang et al. 2010). This suggests that a very large number of SNPs have very small (perhaps additive) effects on this phenotype.

On the opposite side of this continuum lie very rare variants with large effect sizes. This scenario is typical for monogenic disorders, hence the success of family-based linkage studies in identifying such variants. But as argued by Pritchard (2001), rare variants could also play a role in the genetics of complex diseases, with DNA sequencing representing the key technology in this context (Text Box 4.7).

Text Box 4.7. Looking for rare variants of complex traits

Common SNPs (used in genome-wide studies) could be revealing an association between a gene locus and a particular phenotype because of multiple rare variants present in the locus. This seems to be the case, for example, for the melatonin receptor 1B gene (*MTNR1B*) and Type 2 Diabetes (Bonnefond et al. 2012). Here, the previous GWAS studies have revealed a strong association between non-coding variants in *MTNR1B* and Type 2 Diabetes; but the effect size is quite small (Odds Ratio [OR] of ~1.10–1.15). Bonnefond and colleagues sequenced the *MTNR1B* exons in 7,632 individuals (2,186 with diabetes) and found 36 very rare (MAF < 0.1 %), non-synonymous variants associated with the disease, with a much higher effect size (OR $= 3.3$). Once they excluded non-functional mutations (through functional assays), the remaining 26 mutations, which are associated with the partial or complete loss of melatonin binding and/or signalling, contributed to the presence of Type 2 Diabetes at an even higher level (OR $= 5.67$).

Finally, what about the middle portion of the Fig. 4.11: intermediate variants with intermediate effects? There are several possibilities for looking for "missing heritability" in this context. First of all, it is likely that *genes interact with other genes* to give rise to a complex trait. Various approaches are being developed to identify SNP–SNP interactions in GWAS datasets (e.g. Dinu et al. 2012). Second, *genes interact with environment*. There are, of course, multiple examples of such interactions; the one between stress and genetic variations in the serotonin transporter gene—leading to depression—is but a single example (Caspi et al. 2010). The interacting effect of a genetic variation and specific environment can explain a considerable amount of variance. We have shown, for example, that a SNP in the *KCTD8* locus interacts with prenatal exposure to maternal cigarette smoking to explain 22 % of variance in the total cortical area (Paus et al. 2012).

Finally, other genetic (e.g. CNVs, Fig. 4.5) and epigenetic (e.g. methylations, acetylations; see Chap. 5) variations are likely to help in finding the "missing heritability".

In summary, incredible advancements in genetics and genomics have opened up new opportunities for systems-level neuroscientists to narrow the gap between their understanding of the human brain and the molecular events shaping it throughout life.

References

Altmuller J, Palmer LJ, Fischer G, Scherb H, Wjst M (2001) Genomewide scans of complex human diseases: true linkage is hard to find. Am J Hum Genet 69(5):936–950. doi:10.1086/324069

Anderson CA, Pettersson FH, Clarke GM, Cardon LR, Morris AP, Zondervan KT (2010) Data quality control in genetic case-control association studies. Nat Protoc 5(9):1564–1573

Bonnefond A, Clément N, Fawcett K, Yengo L, Vaillant E, Guillaume JL, Dechaume A, Payne F, Roussel R, Czernichow S, Hercberg S, Hadjadj S, Balkau B, Marre M, Lantieri O, Langenberg C, Bouatia-Naji N, Charpentier G, Vaxillaire M, Rocheleau G, Wareham NJ, Sladek R, McCarthy MI, Dina C, Barroso I, Jockers R, Froguel P (2012) Meta-Analysis of Glucose and Insulin-Related Traits Consortium (MAGIC). Rare MTNR1B variants impairing melatonin receptor 1B function contribute to type 2 diabetes. Nat Genet 44(3):297–301
Bouchard TJ Jr, McGue M (2003) Genetic and environmental influences on human psychological differences. J Neurobiol 54(1):4–45. doi:10.1002/neu.10160
Caspi A, Hariri AR, Holmes A, Uher R, Moffitt TE (2010) Genetic sensitivity to the environment: the case of the serotonin transporter gene and its implications for studying complex diseases and traits. Am J Psychiatry 167(5):509–527. doi:10.1176/appi.ajp.2010.09101452
Chen ZY, Patel PD, Sant G, Meng CX, Teng KK, Hempstead BL, Lee FS (2004) Variant brain-derived neurotrophic factor (BDNF) (Met66) alters the intracellular trafficking and activity-dependent secretion of wild-type BDNF in neurosecretory cells and cortical neurons. J Neurosci 24(18):4401–4411
Colhoun HM, McKeigue PM, Davey Smith G (2003) Problems of reporting genetic associations with complex outcomes. Lancet 361(9360):865–872
Collins FS, Patrinos A, Jordan E, Chakravarti A, Gesteland R, Walters L (1998) New goals for the U.S. Human Genome Project: 1998–2003. Science 282:682–689
Conrad DF, Pinto D, Redon R, Feuk L, Gokcumen O, Zhang Y, Aerts J, Andrews TD, Barnes C, Campbell P, Fitzgerald T, Hu M, Ihm CH, Kristiansson K, Macarthur DG, Macdonald JR, Onyiah I, Pang AW, Robson S, Stirrups K, Valsesia A, Walter K, Wei J, Wellcome Trust Case Control Consortium, Tyler-Smith C, Carter NP, Lee C, Scherer SW, Hurles ME (2010) Origins and functional impact of copy number variation in the human genome. Nature 464(7289):704–712
Dinu I, Mahasirimongkol S, Liu Q, Yanai H, Sharaf Eldin N, Kreiter E, Wu X, Jabbari S, Tokunaga K, Yasui Y (2012) SNP–SNP interactions discovered by logic regression explain Crohn's Disease Genetics. PLoS ONE 7(10):e43035. doi:10.1371/journal.pone.0043035
Dudbridge F, Gusnanto A (2008) Estimation of significance thresholds for genomewide association scans. Genet Epidemiol 32(3):227–234. doi:10.1002/gepi.20297
Giedd JN, Schmitt JE, Neale MC (2007) Structural brain magnetic resonance imaging of pediatric twins. Hum Brain Mapp 28:474–481
Gillis J, Pavlidis P (2011) The role of indirect connections in gene networks in predicting function. Bioinformatics 27(13):1860–1866. doi:10.1093/bioinformatics/btr288
Gillis J, Mistry M, Pavlidis P (2010) Gene function analysis in complex data sets using ErmineJ. Nat Protoc 5 (6):1148-1159. doi:nprot.2010.78 [pii] 10.1038/nprot.2010.78
Glahn DC, Thompson PM, Blangero J (2007) Neuroimaging endophenotypes: strategies for finding genes influencing brain structure and function. Hum Brain Mapp 28(6):488–501. doi:10.1002/hbm.20401
Gusella JF, Wexler NS, Conneally PM, Naylor SL, Anderson MA, Tanzi RE, Watkins PC, Ottina K, Wallace MR, Sakaguchi AY et al (1983) A polymorphic DNA marker genetically linked to Huntington's disease. Nature 306(5940):234–238
Igl BW, Konig IR, Ziegler A (2009) What do we mean by 'replication' and 'validation' in genome-wide association studies? Hum Hered 67(1):66–68. doi:10.1159/000164400
International HapMap C, Altshuler DM, Gibbs RA, Peltonen L, Altshuler DM, Gibbs RA, Peltonen L, Dermitzakis E, Schaffner SF, Yu F, Peltonen L, Dermitzakis E, Bonnen PE, Altshuler DM, Gibbs RA, de Bakker PI, Deloukas P, Gabriel SB, Gwilliam R, Hunt S, Inouye M, Jia X, Palotie A, Parkin M, Whittaker P, Yu F, Chang K, Hawes A, Lewis LR, Ren Y, Wheeler D, Gibbs RA, Muzny DM, Barnes C, Darvishi K, Hurles M, Korn JM, Kristiansson K, Lee C, McCarrol SA, Nemesh J, Dermitzakis E, Keinan A, Montgomery SB, Pollack S, Price AL, Soranzo N, Bonnen PE, Gibbs RA, Gonzaga-Jauregui C, Keinan A, Price AL, Yu F, Anttila V, Brodeur W, Daly MJ, Leslie S, McVean G, Moutsianas L, Nguyen H, Schaffner SF, Zhang Q, Ghori MJ, McGinnis R, McLaren W, Pollack S, Price AL, Schaffner SF, Takeuchi F, Grossman SR, Shlyakhter I, Hostetter EB, Sabeti PC, Adebamowo CA, Foster MW,

Gordon DR, Licinio J, Manca MC, Marshall PA, Matsuda I, Ngare D, Wang VO, Reddy D, Rotimi CN, Royal CD, Sharp RR, Zeng C, Brooks LD, McEwen JE (2010) Integrating common and rare genetic variation in diverse human populations. Nature 467 (7311):52–58. doi:10.1038/nature09298

Klopocki E, Mundlos S (2011) Copy-number variations, noncoding sequences, and human phenotypes. Annu Rev Genomics Hum Genet 12:53–72. doi:10.1146/annurev-genom-082410-101404

Kumar P, Henikoff S, Ng PC (2009) Predicting the effects of coding non-synonymous variants on protein function using the SIFT algorithm. Nat Protoc 4(7):1073–1081. doi:10.1038/nprot.2009.86

Lander E, Kruglyak L (1995) Genetic dissection of complex traits: guidelines for interpreting and reporting linkage results. Nat Genet 11:241–247

Lee HK, Braynen W, Keshav K, Pavlidis P (2005) ErmineJ: tool for functional analysis of gene expression data sets. BMC Bioinform 6:269. doi:10.1186/1471-2105-6-269

Levy S, Sutton G, Ng PC, Feuk L, Halpern AL, Walenz BP, Axelrod N, Huang J, Kirkness EF, Denisov G, Lin Y, MacDonald JR, Pang AW, Shago M, Stockwell TB, Tsiamouri A, Bafna V, Bansal V, Kravitz SA, Busam DA, Beeson KY, McIntosh TC, Remington KA, Abril JF, Gill J, Borman J, Rogers YH, Frazier ME, Scherer SW, Strausberg RL, Venter JC (2007) The diploid genome sequence of an individual human. PLoS Biol 5(10):e254. doi:10.1371/journal.pbio.0050254

Paus T, Bernard M, Chakravarty MM, Davey Smith G, Gillis J, Lourdusamy A, Melka MG, Leonard G, Pavlidis P, Perron M, Pike GB, Richer L, Schumann G, Timpson N, Toro R, Veillette S, Pausova Z (2012) KCTD8 gene and brain growth in adverse intrauterine environment: a genome-wide association study. Cereb Cortex 22(11):2634–2642

Pan Q, Shai O, Lee LJ, Frey BJ, Blencowe BJ (2008) Deep surveying of alternative splicing complexity in the human transcriptome by high-throughput sequencing. Nat Genet 40(12):1413–1415. doi:10.1038/ng.259

Pe'er I, Yelensky R, Altshuler D, Daly MJ (2008) Estimation of the multiple testing burden for genomewide association studies of nearly all common variants. Genet Epidemiol 32(4):381–385. doi:10.1002/gepi.20303

Peper JS, Brouwer RM, Boomsma DI, Kahn RS, Hulshoff Pol HE (2007) Genetic influences on human brain structure: a review of brain imaging studies in twins. Hum Brain Mapp 28(6):464–473. doi:10.1002/hbm.20398

Pritchard JK (2001) Are rare variants responsible for susceptibility to complex diseases? Am J Hum Genet 69(1):124–137. doi:10.1086/321272

Russcher H, van Rossum EF, de Jong FH, Brinkmann AO, Lamberts SW, Koper JW (2005) Increased expression of the glucocorticoid receptor-A translational isoform as a result of the ER22/23EK polymorphism. Mol Endocrinol 19(7):1687–1696. doi:10.1210/me.2004-0467

Stenson PD, Mort M, Ball EV, Howells K, Phillips AD, Thomas NS, Cooper DN (2009) The Human Gene Mutation Database: 2008 update. Genome medicine 1(1):13. doi:10.1186/gm13

Strachan T, Read A (2011) Human molecular genetics. Garland Science, 4 edn. Taylor and Francis Group, LLC, New York USA, Abingdon UK

Studies N-NWGoRiA, Chanock SJ, Manolio T, Boehnke M, Boerwinkle E, Hunter DJ, Thomas G, Hirschhorn JN, Abecasis G, Altshuler D, Bailey-Wilson JE, Brooks LD, Cardon LR, Daly M, Donnelly P, Fraumeni JF, Jr., Freimer NB, Gerhard DS, Gunter C, Guttmacher AE, Guyer MS, Harris EL, Hoh J, Hoover R, Kong CA, Merikangas KR, Morton CC, Palmer LJ, Phimister EG, Rice JP, Roberts J, Rotimi C, Tucker MA, Vogan KJ, Wacholder S, Wijsman EM, Winn DM, Collins FS (2007) Replicating genotype-phenotype associations. Nature 447 (7145):655–660. doi:10.1038/447655a

Visscher PM (2008) Sizing up human height variation. Nat Genet 40(5):489–490. doi:10.1038/ng0508-489

Watson JD, Crick FH (1953) Molecular structure of nucleic acids; a structure for deoxyribose nucleic acid. Nature 171(4356):737–738 dbSNP Build 135 Summary. Available via Bethesda (MD): National Center for Biotechnology Information, National Library of Medicine. http://www.ncbi.nlm.nih.gov.myaccess.library.utoronto.ca/SNP/snp_summary_byOrg.cgi?tax_id=9606&build_id=135

Hindorff LA, MacArthur JEBIW A, Junkins HA, Hall PN, Klemm AK, Manolio TA. A Catalog of Published Genome-Wide Association Studies. www.genome.gov/gwastudies

Yang J, Benyamin B, McEvoy BP, Gordon S, Henders AK, Nyholt DR, Madden PA, Heath AC, Martin NG, Montgomery GW, Goddard ME, Visscher PM (2010) Common SNPs explain a large proportion of the heritability for human height. Nat Genet 42(7):565–569. doi:10.1038/ng.608

Zhang L, Zhou R, Li X, Ursano RJ, Li H (2006) Stress-induced change of mitochondria membrane potential regulated by genomic and non-genomic GR signaling: a possible mechanism for hippocampus atrophy in PTSD. Med Hypotheses 66(6):1205–1208. doi:10.1016/j.mehy.2005.11.041

Chapter 5
Epigenomics

Not all genes are expressed in all tissues at all times. While many molecular mechanisms regulating gene expression (in space and over time) are coded in the DNA sequence (e.g. enhancers, repressors, transcription factors), there is a number of so-called epigenetic mechanisms that can regulate gene expression by other means.

In this chapter, we will first review the basics of epigenetics and then describe the two most common epigenetic mechanisms, DNA methylation and histone modification. We will conclude by touching upon a few issues relevant for the integration of genomic and epigenomic information in population-based studies.

5.1 Epigenetics: Heritable, Stochastic and Environment-Induced

In the broad sense, there are three general sources of epigenetic modifications and their variations across cells, tissues and individuals: (1) *heritable* modifications, which can be inherited either from cell to daughter cell (i.e. within the life of an organism) or from a parent to a child (i.e. across generations); (2) modifications that arise from *stochastic instability* in the transfer of epigenetic markers during cell divisions; and (3) epigenetic modifications induced by the *environment*.

5.1.1 Heritable Modifications

Imprinting and X-inactivation are the two most common examples of epigenetic modifications that are inherited from cell to daughter cell (but not from parent to child). In the case of X-inactivation (in females), either the maternal or paternal X chromosome is inactivated in the progenitor cell of a particular lineage (e.g. neuron, oligodendrocyte, hepatocyte). Although the initial "choice" as to which of the two X chromosomes to inactivate is random, all cells derived (and re-derived)

T. Paus, *Population Neuroscience*, DOI: 10.1007/978-3-642-36450-1_5,

subsequently from the original progenitor cell inactivate the same (maternal or paternal) X chromosome. A given X-inactivation (e.g. of the X chromosome inherited from the mother) is inherited through subsequent cell divisions (mitosis) during the life of the individual, but it is not passed on to the offspring; thus, the offspring can inherit an X chromosome that has been either active or inactive in his/her mother. Note that this mechanism creates a mosaic of cells containing either an active or inactive maternal (or paternal) X chromosome.

Similar epigenetic mechanisms, called gene imprinting, may also regulate expression of genes located on autosomal chromosomes. Thus, there are more than a hundred "mono-allelic" genes in which the allele inherited from one parent is silenced (imprinted) while the other one remains active (Henckle and Arnaud 2010). Furthermore, it appears that gene imprinting can operate in an age-related fashion: for example, maternal alleles of certain genes are expressed in the embryonic brain (while the paternal alleles of the same genes are silenced/ imprinted) and vice versa in the adult brain (Gregg et al. 2010).

Trans-generational (from parent to child) transmission of epigenetic modifications has been a surprising discovery. For most of the past 150 years, Darwin's theory of natural selection was assumed to be the only (that is to say, correct) alternative to the view formulated by Jean-Baptiste Lamarck at the beginning of the nineteenth century: that acquired traits can be passed from parent to child (Lamarck 1809). Simply put, we have taken for granted that Lamarckism was wrong and that acquired traits *cannot* be inherited. This view is now changing, however.

Although it is true that most epigenetic marks underlying the relatively stable transmission of epigenetic information from cell to daughter cell during the life of an individual (see above) are erased during gametogenesis,[1] some of these marks can be passed on to the offspring. Thus, we have trans-generational epigenetic inheritance (reviewed in Daxinger and Whitelaw 2012). An often-quoted example of this kind of inheritance is the agouti viable yellow (A^{vy}) epiallele.[2] In the genetically identical (i.e. isogenic) A^{vy} mice, the degree of transcription of a retrotransposon[3] located upstream of the agouti (A) gene results in the various degree of the expression of this gene and, in turn, a varied amount of the agouti protein being deposited in the mouse fur. The latter affects the fur colour on a continuum, starting with the full-yellow (all-cell) coat, going into a mosaic of yellow and agouti hair and ending with the full-agouti coat. Importantly, the level of transcription of this transposable element varies in proportion to its methylation; hence, an epiallele: less methylation means more transcription and more yellow colour (Morgan et al. 1999; see Fig. 5.1). This effect appears to be related to an

[1] Gametogenesis refers to the production of gametes (eggs and sperm) in gonads, through meiosis (see Sect. 4.1.)

[2] Epialleles differ in their epigenetic modifications (whereas alleles differ in nucleotides).

[3] Retrotransposon is a form of the transposable elements that first copy themselves from DNA to RNA (transcription), then back to DNA (reversed transcription), before inserting themselves into the genome in a new position. In this way, they generate insertions, deletions and translocations.

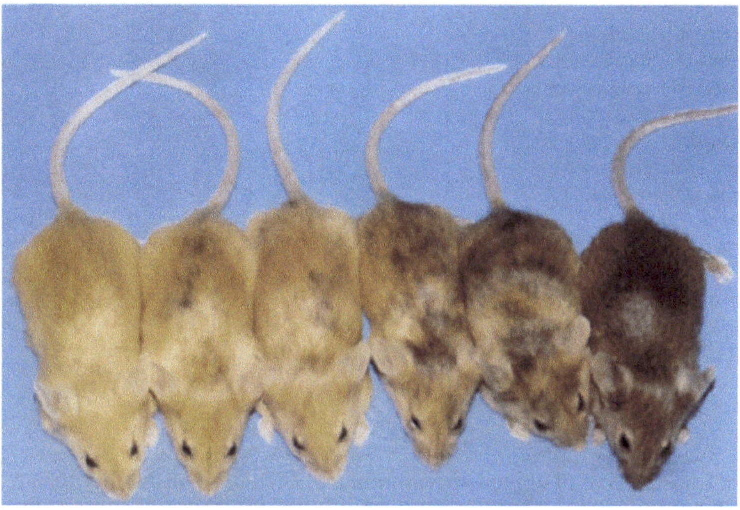

Fig. 5.1 Agouti mice. From Morgan et al. (1999)

incomplete erasure of the epigenetic mark on the retrotransposon in the maternal (but not paternal) germ line.

Such trans-generational epigenetic transmission is likely to vary with the number of subsequent generations. For example, certain epigenetic modifications of histones in the chromatin of *Caenorhabditis elegans* parents, known to affect their longevity, are not entirely erased during gametogenesis and, therefore, can also extend the lifespan of the offspring up to the third generation (Greer et al. 2011). In this case, this trans-generational inheritance appears to affect preferentially the epigenetic regulation (expression) of genes involved in metabolic pathways (Greer et al. 2011).

What are the sources of inter-individual variability in the epigenome? In the next two sections, we will discuss stochastic instability in the transfer of epigenetic markers during mitosis and environment-induced epigenetic modifications (see Fig. 5.2).

5.1.2 Stochastic Instability

Epigenetic marks are transmitted from cell to daughter cell during mitosis and, to a much lesser extent, from a parent to offspring through the germ line. This transfer of epigenetic information involves a variety of molecular processes (see Sect. 5.2 for details) and, not surprisingly, it is prone to errors. Thus, it has been estimated that the error rate for replicating epigenetic marks is ~ 1 in 1,000; this is much higher than the estimated error rate during the DNA replication [~ 1 in 1,000,000 bases; Hjelmeland (2011)]. These errors introduce a certain level of randomness

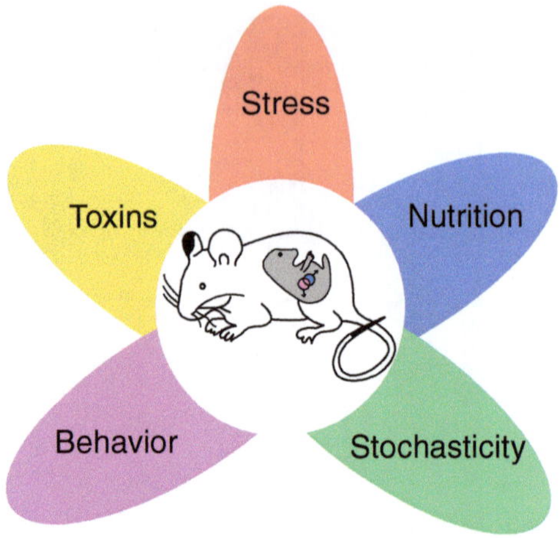

Fig. 5.2 Sources of epigenetic variability. Five environmental influences that affect the developing embryo and its primordial germ cells (represented by the *pink* and *blue dots*). Adapted based on Faulk and Dolinoy (2011)

into the transmission of epigenetic information. Stochastic instability then refers to the probabilistic nature of the processes that results both from a predictable action and the random element. The above example of agouti mice illustrates the range of such stochastic instability: the level of DNA methylation of the "metastable epi-allele" can vary by over 80 % across the isogenic mice; a phenomenon stable during the life of an individual but stochastic across different individuals (Dolinoy et al. 2010). What are the main sources of the predictable actions and the most important time window of their influence on the epigenome?

5.1.3 Environment-Induced Epigenetic Modifications

During gametogenesis, epigenetic marks are first (largely) erased and then re-established *de novo*. Thus, *in utero*, the re-establishment of epigenetic marks can be influenced by a variety of *environmental influences* acting on the pregnant mother (F0 generation) and both the somatic and germ lines of the embryo (F1 generation). In the case of the former, such somatic "epimutations" will be transmitted from cell to daughter cell throughout the prenatal and post-natal life of the exposed (F1) offspring; these effects are likely to be present in all tissues. In the case of the latter, the germ-line epimutations have the potential for being transmitted to the subsequent (F2) generation, and beyond (see Fig. 1.3). On the other hand, if the environment acts *post-natally*, its effect is more likely to be seen in specific tissues; thus, tobacco smoke is more likely to affect lung tissue, and diet (together with local bacteria; see the paragraph on "body environment" in Sect. 3.1) is more likely to affect the gut tissue. Figure 5.2 illustrates the most common exposures that have been investigated in the context of environment-induced changes in epigenome; these vary from the effects of specific diets, toxins, stress and behaviour (reviewed in Faulk and Dolinoy 2011). One of the most exiting models of behaviour-induced epigenetic modifications has been introduced by Michael Meaney and his colleague; in a series of experiments, they showed that high (vs. low) levels of licking and grooming (by the dam) during the first week of pups' lives is associated with lower levels of methylation of the promoter of the glucocorticoid receptor (and, in turn, its higher expression) in the hippocampus, as well as lower response of the hypothalamic-pituitary-adrenal axis to stress (reviewed in Zhang et al. 2013; see Fig. 5.3).

As pointed out above, epigenetic modifications involve a variety of mechanisms, including DNA methylation, modifications of histone proteins and post-transcriptional modifications of non-coding RNA, such as microRNAs, which regulate translation of mRNA into polypeptides. I will now describe two of these mechanisms in some detail.

5.2 DNA Methylation and Histone Modifications

The methylation of cytosine is the most common epigenetic modification of DNA. Given that the diploid human genome contains 6×10^9 nucleotides (A, C, G and T), there are about 150,000,000 cytosines that—in theory—can exist, in either a methylated or unmethylated state. But, in fact, methylation takes place most often

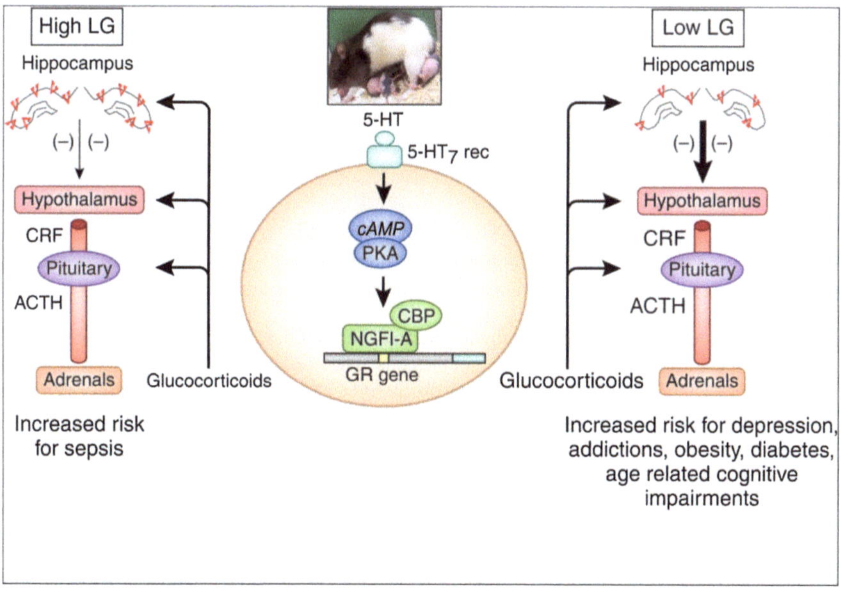

Fig. 5.3 Associations between maternal behaviour (licking and grooming), expression of the glucocorticoid receptor in the hippocampus, regulation of the hypothalamus-pituitary-adrenal axis and psychopathology (*right side*). *LG* licking and grooming, *ACTH* adrenocorticotropin, *CRF* corticotropin releasing factor, *5-HT* serotonin, *cAMP* cyclic adenosine monophosphate, *PKA* protein kinase A, *NGFI-A* nerve growth factor–inducible factor A, *CBP* CREB-binding protein, *GR* glucocorticoid receptor. From Zhang et al. (2013)

(but not exclusively) when cytosine sits next to guanine—that is, if it is present as CpG[4] dinucleotides; when CpG dinucleotides cluster together, we talk about CpG islands. Note that about 60 % of human gene promoters are associated with CpG islands, thus providing a powerful means for DNA methylation to influence gene expression (Portela and Esteller 2010). The so-called CpG shores (areas flanking the CpG islands) show higher variation in DNA methylation, even though the CpG density is lower in these regions as compared with the CpG islands (Rakyan et al. 2011). As shown in Fig. 5.4, DNA methylation can occur not only in CpG islands and shores but also in the gene body, and on repetitive sequences, including those associated with transposable elements (see above for the agouti mice). In general,

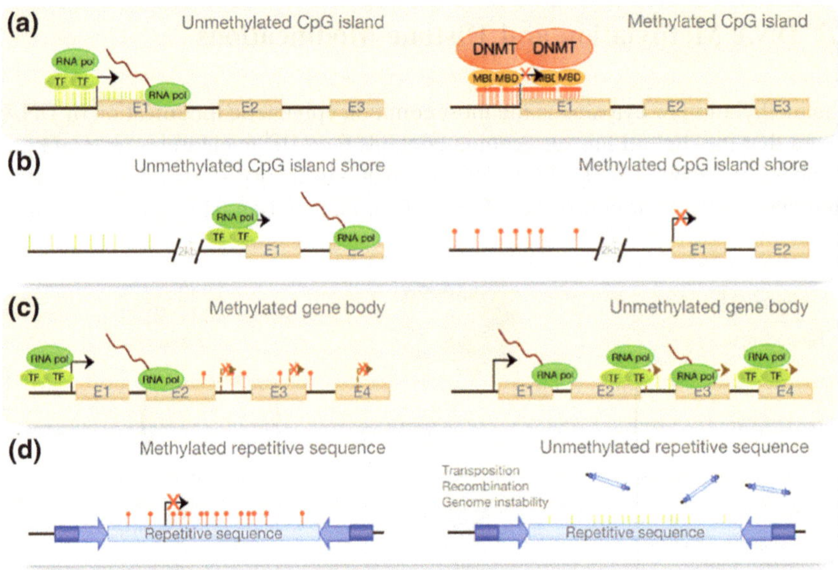

Fig. 5.4 DNA methylation. "DNA methylation can occur in different regions of the genome. The alteration of these patterns leads to disease in the cells. The normal scenario is depicted in the *left column* and alterations of this pattern are shown on the *right*. **a** CpG islands at promoters of genes are normally unmethylated, allowing transcription. Aberrant hypermethylation leads to transcriptional inactivation. **b** The same pattern is observed when studying island shores, which are located up to 2 kb upstream of the CpG island. **c** However, when methylation occurs at the gene body, it facilitates transcription, preventing spurious transcription initiations. In disease, the gene body tends to demethylate, allowing transcription to be initiated at several incorrect sites. **d** Finally, repetitive sequences appear to be hypermethylated, preventing chromosomal instability, translocations and gene disruption through the reactivation of endoparasitic sequences. This pattern is also altered in disease." *E* Exon, *TF* transcription factor, *RNA pol* RNA polymerase, *DNMT* DNA methyltransferase, *MBD* methyl-CpG-binding-domain proteins. From Portela and Esteller (2010)

[4] Note the nomenclature: CG refers to cytosine located on one DNA strand and guanine on the other (complementary) strand. On the other hand, CpG refers to the two nucleotides being located side by side on the same DNA strand.

methylation of the CpG islands at promoters of genes is associated with a reduced rate of DNA transcription. On the other hand, methylation in the gene body facilitates gene expression by preventing spurious initiations of transcription (Portela and Esteller 2010).

How does DNA methylation regulate gene expression? One of the mechanisms involves recruitment of special proteins[5] that go on to recruit histone-modifying and chromatin-remodelling complexes to the methylated site. As explained below, the DNA-chromatin complex is a powerful regulator of gene expression by virtue of "opening" and "closing" DNA for transcription (see Sect. 4.1). Another mechanism involves direct inhibition of transcription by preventing the binding of relevant transcription factors to the promoter (e.g. Kuroda et al. 2009).

Recall that DNA is packaged with protein complexes to form chromatin. Heterochromatin contains tightly packed and inactive DNA, whereas euchromatin contains a stretched out and active DNA molecule ready for transcription (Sect. 4.1). Histone proteins are the main actors in the transformation of heterochromatin to euchromatin and back (see Fig. 4.1). Gene expression requires uncoiling of chromatin fibres, a process guided by H1 histone and its bonds with DNA molecules. Once uncoiled, two turns of DNA molecule are wrapped around an octamer of core histones (H2, H3A, H3B and H4) in the individual nucleosomes; the N-terminal tails of the core histones protrude out of a nucleosome (see Fig. 4.1b). Importantly, various chemical processes[6] can modify amino acids in these histone tails leading, in turn, to the binding of different proteins, affecting local condensation of chromatin and, as such, the level of transcriptional activity at this particular stretch of DNA molecule.[7]

5.3 Bringing Together Genome and Epigenome

Advancements made in mapping DNA variations in the human genome have enabled a GWAS-based search for genetic variants associated with complex traits. In the same manner, we are now in a position to carry out genome-wide scans for epigenetic marks; at present, the relevant technology enables us to do so only for DNA methylations (Rakyan et al. 2011). For example, the Infinium HumanMethylation450 BeadChip Kit allows one to assess methylation state at 485,000 methylation sites across the entire genome, with an average of 17 CpG sites per

[5] Methyl-CpG binding-domain proteins.

[6] Acetylation, methylation, phosphorylation, sumoylation and ubiquitylation.

[7] For example, trimethylation of lysine 27 in H3 histone (H3K27me3) is associated with gene silencing (e.g. Soshnikova and Duboule 2008). As described above, deficiencies in another of the histone-trimethylation complexes (H3K4me3) influence the lifespan of *C. elegans* and that epigenetic modifications (demethylation) of this complex, when located in the vicinity of certain genes, are transmitted across four generations, together with the phenotype. In other words, it influences expression of these genes and longevity (Greer et al. 2011).

gene region and coverage of 96 % of CpG islands. In principle, this information can be interrogated in three ways that reflect the combination of time of exposure to the presumed methylation event (prenatal vs. post-natal) and type of cells (somatic or germ lines) affected by the event. This is in addition to the genetic control of methylation events.

In the case of *prenatal events*, such as exposure to cigarette smoking or a diet low on folic acid, we would expect to find epimutations in both somatic and germ lines of the offspring.

In case of *somatic-cell lines*, epimutations should be present across a variety of tissues. Given the limited availability of tissue in human population-based studies, this assumption can be tested—for example—by comparing the epigenome of DNA extracted from blood cells, buccal (epithelial) cells, bulge cells of the hair follicles or epithelial cells found in urine and faeces (Rakyan et al. 2011). Naturally, the global genome-wide rate of epimutations (defined as a deviation from the reference sample) may serve as a useful phenotype and, perhaps, a more precise proxy of the actual level of exposure to the environment under study. Furthermore, the presence of epimutations in a particular gene region provides an important parameter to be used in evaluating interactions between a particular environment (e.g. cigarette smoking during pregnancy) and specific genetic variations (Text Box 5.1).

Text Box 5.1. Smoking during pregnancy and DNA methylation

In one of our studies, we have observed an interaction between the *BDNF* genotype and prenatal exposure to maternal smoking with regard to the relationship between brain and behaviour; but only non-exposed individuals showed the effect of genotype (Lotfipour et al. 2009). We speculated that this gene might have been "silenced" in the exposed offspring, thus rendering DNA variations in the gene irrelevant in this group. We followed up this idea and showed that the two groups (exposed and non-exposed adolescents) differed in the methylation rate in the CpG island of one of the promoters of the *BDNF* gene (Toledo-Rodriguez et al. 2010). Obviously, the methylation rate was assessed here using DNA extracted from blood cells (and not the brain); as pointed out above, however, the exposure presumably associated with the higher methylation rate occurred, in this case, prenatally and, as such, was likely to affect all tissues to a more similar extent than one would expect for post-natal exposure to the same environment.

In case of *germ lines*, epimutations can be transmitted from parent to child. But note that—in the case of prenatal exposure—this process can be a mix of the direct effect of exposure *and* the trans-generational transmission of epigenetic marks. Thus, a given prenatal exposure can be transmitted from the first generation (F0 = smoking mother) to the grandchild (F2) simply by exposing the germ-line cells of the child (F1) and not erasing completely this epigenetic modification at the time of fertilization of the "exposed" (F1) gametes (i.e. oocytes) by non-exposed ones (i.e. sperm). The pure exposure-independent trans-generational transmission of the epigenetic information thus takes place only in subsequent generations (F2–F3, F3–F4, etc.). Thus, we need to be cautious when interpreting findings that are based on three generations only (i.e. children of mothers who were exposed prenatally to a given environment).[8]

In summary, we need to consider the following four elements of trans-generational transmission: (1) epimutations of somatic-cell lines of the F1 embryo/foetus; (2) epimutations of germ lines of the F1 embryo; (3) incomplete erasure of epigenetic marks during fertilization, giving rise to the F2 generation; and (4) transmission of epigenetic marks from F2 to F3 (and subsequent) generations.

Finally, in case of post-natal exposures, most of such epigenetic effects are likely to affect only tissues that either are in direct physical contact with the agent (e.g. cigarette smoke) or represent a target tissue in a biochemical cascade initiated by the agent (e.g. stress-induced stimulation of glucorticoid receptors of the hippocampus). For the human brain, such tissue-specific epigenetic modifications can be assessed only via surgical removals or biopsies (e.g. a repository for gliomas [http://caintegrator.nci.nih.gov/rembrandt/]; Riddick and Fine 2011) or in post-mortem tissue (e.g. McGowan et al. 2009).

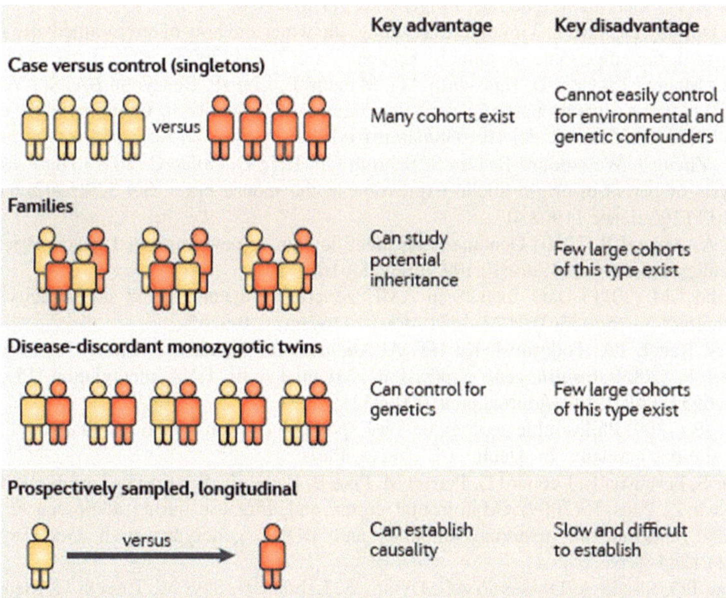

Fig. 5.5 Design of genome-wide epigenetic studies. From Rakyan et al. (2011)

---

[8] The often-quoted examples of a trans-generational effect of food availability on reproduction, health and mortality, such as the Dutch Hunger Winter (Roseboom et al. 2011) and the Overkalix cohort in Northern Sweden (Pembrey et al. 2006), involve only three generations.

Finally, we can ask what are the possible designs of large, genome-wide epigenetic studies. As illustrated in Fig. 5.5, these studies fall into two different classes. In the case of disease-based phenotypes known at the time of the study, one can use either case-control cohorts of unrelated individuals or monozygotic twins discordant for the presence of a given disease. The latter design has the obvious advantage of effectively removing genetic variations as the possible "cause" of the disease through monozygosity. If studying quantitative traits rather than diseases, one can resort either to a family-based design or a longitudinal study. In both cases, one may be able to distinguish between inherited and *de novo* epimutations, especially if biospecimens are available in multiple generations, and evaluate their associations with the system-level phenotype (and genotype) of interest.

References

Daxinger L, Whitelaw E (2012) Understanding transgenerational epigenetic inheritance via the gametes in mammals. Nat Rev Genet 13(3):153–162

Dolinoy DC, Weinhouse C, Jones TR, Rozek LS, Jirtle RL (2010) Variable histone modifications at the A(vy) metastable epiallele. Epigenetics 5(7):637–644

Faulk C, Dolinoy DC (2011) Timing is everything: the when and how of environmentally induced changes in the epigenome of animals. Epigenetics 6(7):791–797

Greer EL, Maures TJ, Ucar D, Hauswirth AG, Mancini E, Lim JP, Benayoun BA, Shi Y, Brunet A (2011) Transgenerational epigenetic inheritance of longevity in Caenorhabditis elegans. Nat 479(7373):365–371. doi:10.1038/nature10572

Gregg C, Zhang J, Weissbourd B, Luo S, Schroth GP, Haig D, Dulac C (2010) High-resolution analysis of parent-of-origin allelic expression in the mouse brain. Sci 329(5992):643–648. doi:10.1126/science.1190830

Henckel A, Arnaud P (2010) Genome-wide identification of new imprinted genes. Brief Funct Genomics 9(4):304–314. doi:10.1093/bfgp/elq016

Hjelmeland LM (2011) Dark matters in AMD genetics: epigenetics and stochasticity. Invest Ophthalmol Vis Sci 52(3):1622–1631. doi:10.1167/iovs.10-6765

Kuroda A, Rauch TA, Todorov I, Ku HT, Al-Abdullah IH, Kandeel F, Mullen Y, Pfeifer GP, Ferreri K (2009) Insulin gene expression is regulated by DNA methylation. PLoS One 4(9):e6953. doi:10.1371/journal.pone.0006953

Lamarck JB (1809) Philosophie zoologique: ou Exposition des considérations relative à l'histoire naturelle des animaux. In. Dentu et L'Auteur, Paris

Lotfipour S, Ferguson E, Leonard G, Perron M, Pike B, Richer L, Seguin JR, Toro R, Veillette S, Pausova Z, Paus T (2009) Orbitofrontal cortex and drug use during adolescence: role of prenatal exposure to maternal smoking and BDNF genotype. Arch Gen Psychiatry 66(11):1244–1252

McGowan PO, Sasaki A, D'Alessio AC, Dymov S, Labonte B, Szyf M, Turecki G, Meaney MJ (2009) Epigenetic regulation of the glucocorticoid receptor in human brain associates with childhood abuse. Nat Neurosci 12(3):342–348. doi:10.1038/nn.2270

Morgan HD, Sutherland HG, Martin DI, Whitelaw E (1999) Epigenetic inheritance at the agouti locus in the mouse. Nat Genet 23(3):314–318. doi:10.1038/15490

Pembrey ME, Bygren LO, Kaati G, Edvinsson S, Northstone K, Sjostrom M, Golding J (2006) Sex-specific, male-line transgenerational responses in humans. Eur J Hum Genet 14(2):159–166. doi:5201538 [pii] 10.1038/sj.ejhg.5201538

Portela A, Esteller M (2010) Epigenetic modifications and human disease. Nat Biotechnol 28(10):1057–1068. doi:10.1038/nbt.1685

Rakyan VK, Down TA, Balding DJ, Beck S (2011) Epigenome-wide association studies for common human diseases. Nat Rev Genet 12(8):529–541. doi:10.1038/nrg3000

Riddick G, Fine HA (2011) Integration and analysis of genome-scale data from gliomas. Nat rev Neurol 7(8):439–450. doi:10.1038/nrneurol.2011.100

Roseboom TJ, Painter RC, van Abeelen AF, Veenendaal MV, de Rooij SR (2011) Hungry in the womb: what are the consequences? Lessons from the Dutch famine. Maturitas 70(2):141–145. doi:10.1016/j.maturitas.2011.06.017

Soshnikova N, Duboule D (2008) Epigenetic regulation of Hox gene activation: the waltz of methyls. BioEssays News Rev Mol Cell Dev Bio 30(3):199–202. doi:10.1002/bies.20724

Toledo-Rodriguez M, Lotfipour S, Leonard G, Perron M, Richer L, Veillette S, Pausova Z, Paus T (2010) Maternal smoking during pregnancy is associated with epigenetic modifications of the brain-derived neurotrophic factor-6 exon in adolescent offspring. Am J Med Genet B Neuro psychiatr Genet

Zhang TY, Labonte B, Wen XL, Turecki G, Meaney MJ (2013) Epigenetic mechanisms for the early environmental regulation of hippocampal glucocorticoid receptor gene expression in rodents and humans. Neuro Psycho Pharmacol 38(1):111–123. doi:10.1038/npp.2012.149

References

Chapter 6
Molecular Phenomics

The path leading from a gene to system-level brain phenotypes (see Chap. 7) starts with its transcription to a messenger RNA (gene expression) and proceeds to translation of the mRNA to a protein. In this chapter, we review the basics of the "omics" approaches not previously covered, namely those that generate transcriptome, proteome, metabolome and lipidome, all providing a rich array of *molecular phenotypes* (Fig. 6.1).

6.1 Transcriptomics

Although all somatic cells contain the same genes, different cells express only a subset of these genes at any given time. As discussed in the previous chapter, there are several genetic and epigenetic mechanisms that regulate gene expression, thereby creating a vast variability across individuals in the level of a particular mRNA in a given tissue. This variability can be captured—at a genome-wide level—using RNA microarrays or next-generation sequencing of RNA (RNA-Seq[1]). But given the limited access to brain tissue, how can we use transcriptomics in population-based studies of the human brain? With certain limitations, there are at least four possibilities.

First, we can use transcriptomics when evaluating the functional consequences of a genetic variation on gene expression in non-brain tissues. Is a SNP discovered, for example, in a GWAS of a particular brain phenotype also associated with variations in gene expression in the same cohort? As we have learned in Chap. 5, this would be more likely if the SNP is non-synonymous and/or located in the promoter region of the gene in question.[2] Therefore, it would be helpful to know whether individuals who differ in this particular genotype also differ in the level of gene expression. This can be checked, provided that investigators collected and

[1] Also called "Whole Transcriptome Shotgun Sequencing."

[2] But note that a synonymous SNP located in any region close to a gene may be in linkage disequilibrium with such a "functional" SNP or could influence its transcription by affecting chromatin structure.

T. Paus, *Population Neuroscience*, DOI: 10.1007/978-3-642-36450-1_6,

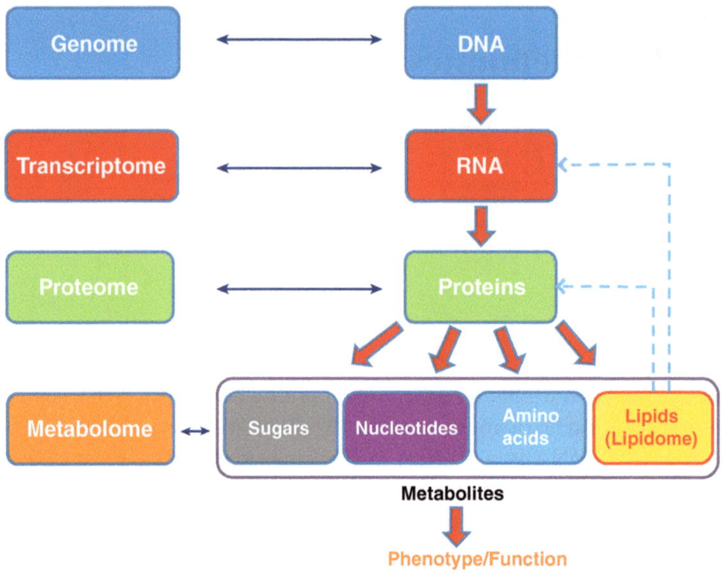

Fig. 6.1 Genome, Transcriptome, Proteome, Metabolome and Lipidome (from http://en.wikipedia.org/wiki/Lipidomics). **Transcriptome** refers to the collection of all genes expressed in a given tissue at a given time. Similarly, a **proteome** is a collection of all proteins produced by an organism at a given time. Finally, proteins exert their action in specific biological pathways (e.g. as enzymes or receptors). A **metabolome** is a collection of thousands of biological molecules used (as substrate) and created (as product) by such pathways. The main classes of these molecules are amino acids, sugars and lipids, the latter constituting a **lipidome**

stored mononuclear leucocytes (lymphocytes and monocytes) and calculated the number of different blood cells per unit of whole blood; 1 ml of blood should contain $\sim 4 \times 10^6$ leucocytes (white blood cells) of which $\sim 1 \times 10^6$ are mononuclear leucocytes. If yes, one can measure the amount of RNA (of the gene of interest) in monocytes and normalize this value by the number of monocytes per unit of volume. Thus, if we find a difference in the level of gene expression as a function of a SNP, we can infer that a genetic variation present at this genomic location (or one that is in linkage disequilibrium) has a potential to influence the transcription of DNA coded there. Of course, we do not know whether this is the case in other tissues, including the brain.

We can use a similar approach in other situations. For example, let us assume that we have uncovered an association between a *known* functional polymorphism—say, resulting in val66met in the BDNF protein (Text Box 4.4)—and a particular brain phenotype in one group of individuals but not in another. While there might be many possible explanations for such a group-by-genotype interaction, one of them could be that *BDNF* is less expressed in the group in which we did observed the *BDNF* effect, thus making the genotype less relevant. Such a scenario may be worth exploring, especially if one suspects environment-induced changes in DNA methylation or histone modification (see Chap. 5).

Second, a number of efforts are underway to construct transcriptomes in the human brain. For example, Kang et al. (2011) reported a number of global sex- and age-related differences in gene expressions assessed with RNA microarrays across 1,340 tissue samples, which were obtained from 16+ brain regions in 57 individuals, from embryos and foetuses, through children and adolescents to adults. The Allen Institute for Brain Science has released the first version of the Human Brain Atlas, containing relative values of gene expression for over a thousand distinct brain regions assessed with RNA microarray in a mere three individuals. These values are plotted in an atlas aligned with the MNI152 target brain, thus allowing 3D localization of each sample in the standardized sterotaxic space used in the majority of structural and functional imaging studies (Fig. 6.2). The two examples illustrate the enormity of constructing a representative transcriptome of the human brain. On the one hand (Kang et al. 2011), we have transcriptomic data obtained in a handful of brain regions in a number of individuals who differ widely in age (from 4 post-conception weeks to 60 years). On the other hand (The Allen Human Brain Atlas), we have transcriptomic data in over 1,000 brain locations mapped in sterotaxic space in three adult brains. The future will show whether such efforts can be expanded to provide a robust (probabilistic) representation of gene expression in the human brain during key developmental periods.

Third, we can use transcriptomes deposited in existing gene-expression databases to evaluate co-expression networks in the relevant tissues (e.g. Gillis et al. 2010; Gillis and Pavlides 2011). This type of analysis relies on evaluating profiles of co-expression between the target gene—revealed, for example, by a GWAS—

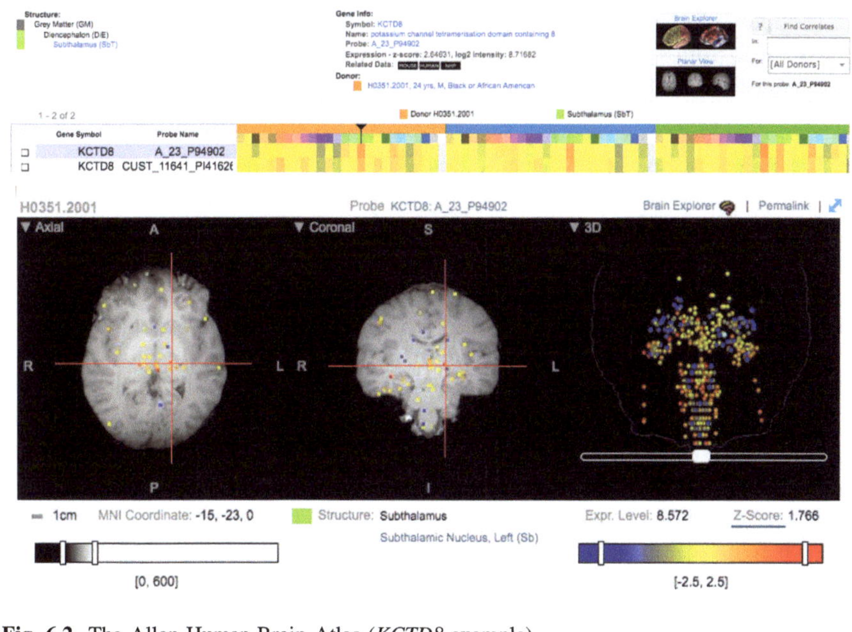

Fig. 6.2 The Allen Human Brain Atlas (*KCTD8* example)

and all other genes included in RNA microarrays collected across hundreds of experiments carried out in different laboratories. In this way, one can reveal biological pathways relevant for the trait in question and, in turn, explore the possible associations between the particular phenotype and any DNA variations in genes co-expressed with the target gene. This approach is also helpful in situations when the gene revealed by a GWAS does not have a well-established function; identifying the gene's co-expression network provides additional information about the possible biological pathways relevant for a given genotype–phenotype relationship (Text Box 6.1.).

Text Box 6.1. Co-expression networks

We found that genetic variations in the *KCTD8* locus are associated with the total brain volume in adolescent girls (Paus et al. 2012). Given that the function of this gene is largely unknown, we used the co-expression approach to identify biological pathways in which this gene might play a role. By interrogating over 600 microarrays of gene expression in foetal-mice tissue, we found that genes co-expressed with *KCTD8* include those playing a role in potassium ion transport (*KCNC3*), development (e.g. *NRG3*, *NRCAM*) and neurotransmission (e.g. *Slc6a13*, *Slc7a10*, *Slc6a3*). We also analysed a small set of microarrays obtained in four human foetal brains (Johnson et al. 2009) and found that genes co-expressed with *KCTD8* include those linked to brain-related gene ontologies, such as "neuron fate commitment," "neural crest cell migration" and "neurotransmitter metabolic process" (Paus et al. 2012; http://www.chibi.ubc.ca/Paus). These two examples illustrate how co-expression analyses carried out with RNA data acquired from relevant tissues may facilitate the interpretation of novel findings obtained in population-based studies.

The fourth and perhaps most exciting possibility for transcriptomics vis-à-vis the human brain is that of converting mature somatic cells (obtained from peripheral tissues like skin) into neuronal cells and subsequently studying the pattern of gene expression in these converted cells. Such a reprogramming of the cell genome can be achieved *indirectly*, by converting, for example, skin fibroblasts into induced pluripotency stem (iPS) cells and then differentiating them into neurons (e.g. Abeliovich and Doege 2009), or *directly*, by converting human skin fibroblasts to the so-called human-induced neuronal (hiN) cells (Qiang et al. 2011). Using the latter approach, the pattern of gene expression of mature neurons selected from the hiN cell culture is consistent with a neuronal phenotype, since the expressed genes are enriched in such gene ontologies as "axonal projection" and "neuronal differentiation" (Qiang et al. 2011). Furthermore, genes expressed by such selected hiN cells cluster together with genes expressed in real neurons, obtained post-mortem from brain tissue, rather than with fibroblasts, astrocytes, neural progenitor, pluripotent embryonic stem cells or induced pluripotent stem cells (Qiang et al. 2011).

Overall, the transcriptome provides a natural bridge between the "static" *genome* and "dynamic" molecular and system-level *phenomes*. The main challenge of transcriptomics vis-à-vis the human brain is that of *tissue access,* which is limited to post-mortem tissue, biopsies or surgical removals [brain tumours (Mischel et al. 2004) or removals of epileptogenic tissue (Crino and Becker 2006)] and conversion of non-neuronal cells (e.g. skin fibroblast) to induced neuronal cells. Only the latter approach would be suitable for population-based studies.

6.2 Proteomics, Metabolomics and Lipidomics

Inter-individual variability in the transcriptome translates into the variability in the *proteome*, a collection of proteins present in the human body at a given time. In theory, with ~21,000 protein-coding genes (Lander 2011) and ~7 alternative-splicing events per a multi-exon human gene (Pan et al. 2008), there might be up to 140,000 proteins differing in structure and/or activity (across the various human tissues and throughout a life span). These proteins serve a multitude of functions, from being the building blocks of cellular membranes and the cytoskeleton to acting as hormones (e.g. thyroxin or insulin) and enzymes. The latter catalyse a wide variety of chemical reactions by converting substrates (e.g. cholesterol) to products (e.g. steroid hormones).

The *metabolome* is a collection of substrates and products implicated in the vast number of biological pathways in which proteins play various roles, including those as enzymes, receptors, transporters or convertors of chemical energy to mechanical energy (e.g. motor proteins). These biological processes also include lipid metabolism; as illustrated in Fig. 6.1, the *lipidome* can be seen as a special section of the metabolome.

In the context of population neuroscience, blood is our primary source of tissue for *omics*-based analyses. By definition, proteomics focuses on proteins, which are mostly *very large* molecules (tens of thousands of daltons[3] in weight). On the other hand, metabolomics deals with molecules of various sizes: small (<100 daltons: e.g. minerals), middle-sized (100–1,000 daltons: vitamins, fatty acids, monosaccharides, amino acids, nucleotides) and large (>1,000 daltons: small proteins [e.g. insulin, 5,700 daltons], polysaccharides, lipids). These molecules can be identified and quantified using a number of high-throughput methodologies, which are based mostly on the combination of liquid chromatography (LC; mixture separation) and

[3] Dalton (Da) is the standard unit used for indicating atomic mass. One dalton is approximately equal to the mass of one proton or neutron and has a value of $\sim 1.66 \times 10^{-27}$ kg.

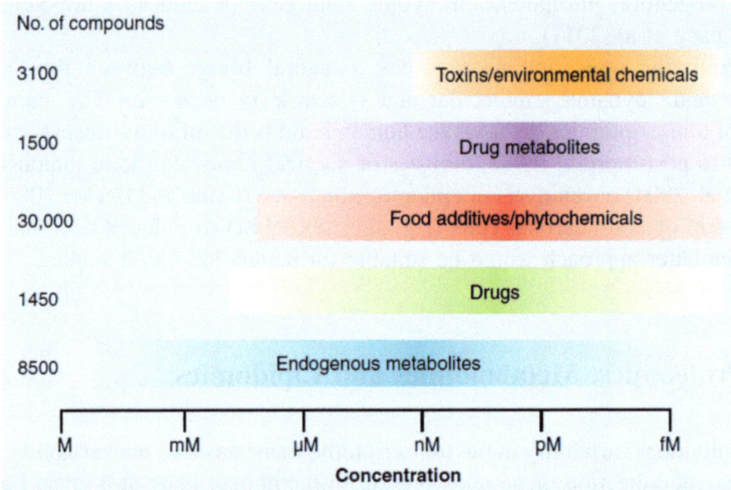

Fig. 6.3 Human metabolome. From Wishart (2011)

mass spectrometry (MS; compound identification), but include also nuclear magnetic resonance (NMR) spectroscopy (Wishart 2008).

Figure 6.3 shows the different components of the human metabolome. It has been estimated that there are ~8,500 "endogenous" compounds (~40,000 if exogenous compounds [drugs, foods and pollutants] are also included), and more than 100,000 if lipids and food metabolites are added as well (Wishart 2011). Note that the concentration of most of the compounds is below 1 μM, requiring measurement through the use of the highly sensitive mass spectrometry.

If we take a blood sample at the time of the system-level assessment, the knowledge of metabolomic and lipidomic profiles provides a rich snapshot of the molecular environment to which the brain has been "exposed" around this time window. In this manner, molecular phenotypes derived with proteomics, metabolomics and lipidomics allow one to explore specific biological pathways as possible mediators of genotype–phenotype relationships or of gene-environment interactions observed at the system level. This issue will be explored in Chap. 10.

Overall, molecular phenotypes—from gene expression to the presence of endogenous and exogenous metabolites—provide complementary knowledge that will facilitate our quest to understand how genes and environment shape the brain and behaviour. The next chapter focuses on tools and concepts available to us for the system-level phenomics of the human brain.

References

Abeliovich A, Doege CA (2009) Reprogramming therapeutics: iPS cell prospects for neurodegenerative disease. Neuron 61(3):337–339. doi:10.1016/j.neuron.2009.01.024 S0896-6273(09)00091-9 [pii]

Crino PB, Becker AJ (2006) Gene profiling in temporal lobe epilepsy tissue and dysplastic lesions. Epilepsia 47(10):1608–1616. doi:10.1111/j.1528-1167.2006.00712.x EPI712 [pii]

Gillis J, Mistry M, Pavlidis P (2010) Gene function analysis in complex data sets using ErmineJ. Nat Protoc 5(6):1148–1159. doi:10.1038/nprot.2010.78 nprot.2010.78 [pii]

Gillis J, Pavlidis P (2011) The role of indirect connections in gene networks in predicting function. Bioinformatics 27(13):1860–1866. doi:10.1093/bioinformatics/btr288

Johnson MB, Kawasawa YI, Mason CE, Krsnik Z, Coppola G, Bogdanovic D, Geschwind DH, Mane SM, State MW, Sestan N (2009) Functional and evolutionary insights into human brain development through global transcriptome analysis. Neuron 62(4):494–509. doi:10.1016/j.neuron.2009.03.027 S0896-6273(09)00286-4 [pii]

Kang HJ, Kawasawa YI, Cheng F, Zhu Y, Xu X, Li M, Sousa AM, Pletikos M, Meyer KA, Sedmak G, Guennel T, Shin Y, Johnson MB, Krsnik Z, Mayer S, Fertuzinhos S, Umlauf S, Lisgo SN, Vortmeyer A, Weinberger DR, Mane S, Hyde TM, Huttner A, Reimers M, Kleinman JE, Sestan N (2011) Spatio-temporal transcriptome of the human brain. Nature 478(7370):483–489. doi:10.1038/nature10523 nature10523 [pii]

Mischel PS, Cloughesy TF, Nelson SF (2004) DNA-microarray analysis of brain cancer: molecular classification for therapy. Nat Rev Neurosci 5(10):782–792. doi:10.1038/nrn1518 nrn1518 [pii]

Qiang L, Fujita R, Yamashita T, Angulo S, Rhinn H, Rhee D, Doege C, Chau L, Aubry L, Vanti WB, Moreno H, Abeliovich A (2011) Directed conversion of Alzheimer's disease patient skin fibroblasts into functional neurons. Cell 146(3):359–371. doi:10.1016/j.cell.2011.07.007 S0092-8674(11)00764-1 [pii]

Wishart DS (2008) Applications of metabolomics in drug discovery and development. Drugs R&D 9(5):307–322

Wishart DS (2011) Advances in metabolite identification. Bioanalysis 3(15):1769–1782. doi:10.4155/bio.11.155

References

Chapter 7
Systems Phenomics

The molecular machinery described in the preceding three chapters represents both the blueprint (genes) and the mechanics (epigenetics, transcriptomics, proteomics) for the emergence of systems-level phenotypes: the functional and structural properties of the human brain. As with any other biological system in our bodies, brain phenotypes arise from a combination of genetic and environmental influences, which are expressed via multiple molecular pathways throughout our lives. Ultimately, inter-individual variability in these structural and functional brain properties underlies differences between individuals in their behavioural traits.[1]

In this chapter, we will focus on the use of magnetic resonance imaging (MRI) for quantifying brain phenotypes. We will start with a few general comments about functional and structural brain imaging, explain the basic principles of MRI, then move on to describe various ways in which we can measure structural and functional brain phenotypes. Given our interest in the ultimate "product" of our brains, namely behaviour, we will also provide a brief overview of tools suitable for out-of-scanner assessment of cognitive abilities and mental health in population-based imaging studies.

7.1 Functional and Structural MRI: An Overview

MRI is a remarkably flexible tool for measuring a wide array of functional and structural properties of the human brain (and other organs). The widespread availability of MR scanners in both research and clinical settings makes it well suited for population-based studies (see Chap. 8). Its non-invasive nature allows one to use MRI as a phenotyping tool throughout the lifespan, from foetal imaging (Fig. 7.1), to that of infants, children and adolescents and, of course, adults.

The same scanner can be used for acquiring different types of images by changing the exact parameters of an MR "scan": the temporal sequence of radiofrequency (RF) pulses, gradients and read-outs (see Sect. 7.2 for basic

[1] We use the term "behaviour" in a broad sense, encompassing not only observable actions but also cognition, emotions and mental health.

T. Paus, *Population Neuroscience*, DOI: 10.1007/978-3-642-36450-1_7,

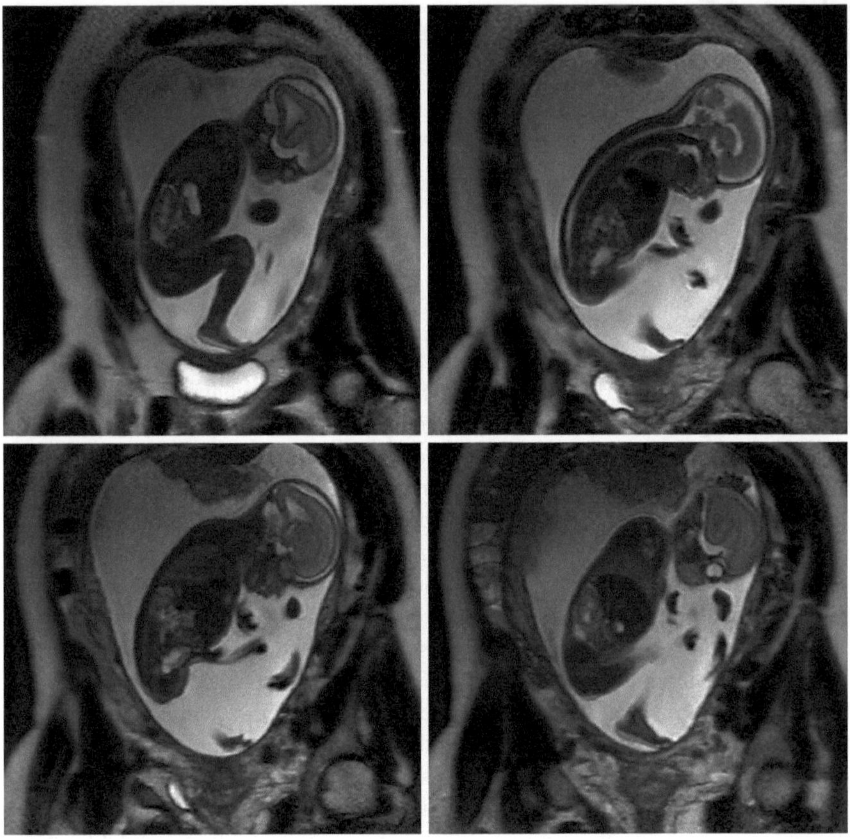

Fig. 7.1 Magnetic resonance images of a 24-week-old foetus. From Anblagan et al. (unpublished observation)

principles). Thus, in one MRI session, we can run several scans to assess a number of functional and structural properties of the human brain (Table 7.1).

Most of these measurements are made evenly throughout the brain—one measurement (or value) per "voxel".[2] Depending on the spatial resolution of the 3D image, thousands of measurements are acquired simultaneously during each scan. Thus, for example, a typical T1-weighted image of the head contains ~10 million voxels (1 × 1 × 1 mm in size), of which about 1.5 million will be inside the brain. But not all MR protocols cover the brain in such an even and unbiased fashion. One notable exception to this desirable feature is that of a paradigm-based functional MRI (fMRI). Although the whole brain is being scanned, we can draw inferences about brain function only from the measurements

[2] Voxel (or volumetric pixel) is a volume element of the (3D) MR image (pixels refer to picture elements of (2D) video images).

Table 7.1 An example of a 60-min MR protocol enabling one to characterize a number of structural and functional properties of the human brain

MRI sequence	Time (min)	Structure and physiology
T1-weighted	10	Volumes, thickness, folding, shape, tissue density
T2-weighted	4	White-matter hyperintensities (number, volume, location)
Diffusion tensor imaging	12	Fractional anisotropy, mean diffusivity, track delineation
Magnetization transfer	8	Myelination index
Arterial spin labelling	5	Perfusion
Resting-state functional	8	Spontaneous cerebral networks; functional connectivity
Paradigm-based functional	6–10	Brain response associated with specific stimuli/tasks; functional connectivity

acquired in the parts of the brain that are engaged by the paradigm (see Sect. 7.4 for details).

As indicated in Table 7.1, MR allows us to assess a large number of structural properties and, unlike fMRI, do so in an unbiased and highly reliable way throughout the brain. Nonetheless, two important issues should be kept in mind when interpreting these measurements: (1) neurobiology underlying the MR metrics and (2) directionality of the brain-behaviour relationships.

First, mapping of the various MR-derived metrics onto the underlying neurobiology is complicated by the fact that each MR voxel contains a mix of various cellular elements (Fig. 1.5). Therefore, we cannot attribute unequivocally a particular biological process to a difference/change in an MR-derived measurement (e.g. cortical thickness) without additional evidence. Genetics may be quite useful in this context. For example, moderating effects of functional polymorphisms in the gene for vascular endothelial growth factor (VEGF) may support one particular interpretation of the observed difference/change in cortical thickness, namely that of differential vascularization. But—of course—only a combination of in vivo (MRI) and ex vivo (e.g. immunohistochemistry) studies in experimental animals can provide more definite answers with regard to the biology underlying MR-based phenotypes.

Second, structure–function relationships can work in both directions: a structural change can precede a functional change and vice versa. It is quite likely, for example, that prenatal and early post-natal events would have significant (and direct) impact on brain growth, with direct consequences on brain function and mental health later in life. We have shown, for example, that an exposure to maternal cigarettes smoking during pregnancy in female offspring (i.e. daughters of the smoking mothers) with a particular genetic variant is associated with a significant ($\sim$10 %) reduction in the overall cortical area (Fig. 7.2), possibly through an accentuated apoptosis of progenitor cells during embryogenesis (Paus et al. 2012). In the long term, this reduction in "brain reserve" might—for example—increase the risk of these individuals for dementia.

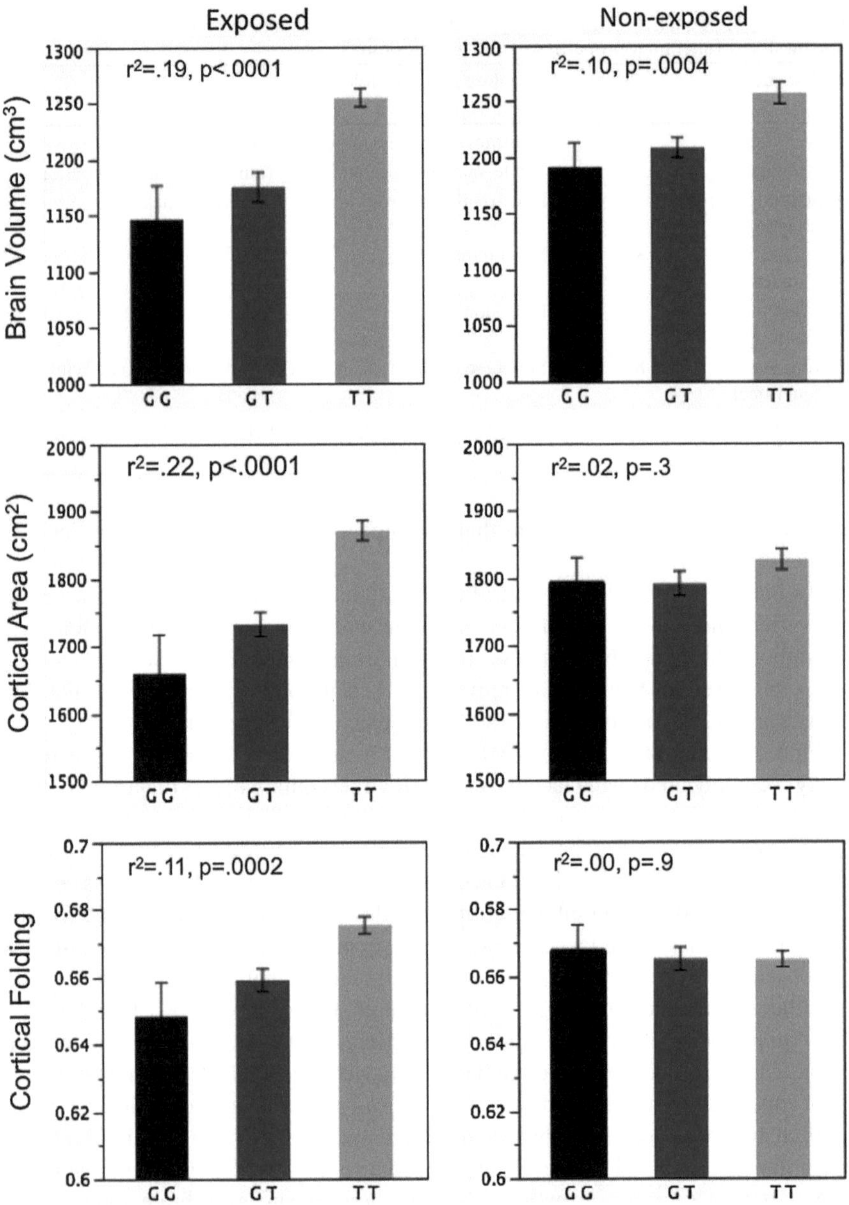

Fig. 7.2 Brain volume, cortical area and cortical folding in female offspring exposed (*left*) and non-exposed (*right*) to maternal cigarette smoking during pregnancy, as a function of *KCTD*8 genotype (rs716890). From Paus et al. (2012)

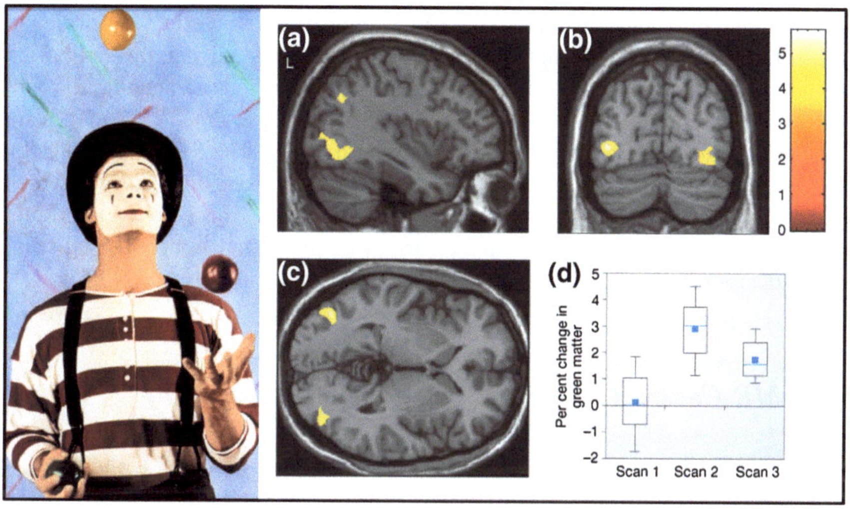

Fig. 7.3 Juggling-related increases in cortical *grey* matter. From Draganski et al. (2004)

On the other hand, a functional engagement of a given brain region may—over time—lead to subtle structural changes detectable with MRI. A classical example of this function-to-structure path, revealed with structural MRI, is that of a repeated practice of juggling: an increase in regional amount of grey matter (GM) in a cortical area engaged by visual motion, after 3 months of learning how to juggle three balls (Fig. 7.3). Note, however, that we can only speculate about the neurobiology (dendrites, vascularization, glia) underlying this change.

Overall, MRI is a highly versatile tool that allows us to quantify a large number of functional and structural brain phenotypes in both global and regional manner. Let us now review the basic principles of this technology.

7.2 MRI: Basic Principles

Nuclei that have an odd number of nucleons (protons and neutrons) possess both a magnetic moment and angular momentum (or spin). In the presence of an external magnetic field, such nuclei "line up" with the main magnetic field and precess (or spin) around their axis at a rate proportional to the strength of the magnetic field, emitting electromagnetic energy as they do. The hydrogen atom contains only a single proton and therefore precesses when exposed to a magnetic field.

In the majority of MR imaging studies, precessing nuclei of hydrogen associated with water and fat is the source of the signal. The signal is generated and measured in the following way. First, the person is exposed to a large static magnetic field (B_0) that aligns hydrogen nuclei along the direction of the applied field. In clinical scanners, the strength of B_0 is most often 1.5 Tesla;[3] B_0 is oriented horizontally, pointing from head to toe along the long axis of the cylindrical magnet. Second, a pulse of electromagnetic energy is applied at a specific RF, with an RF coil placed around (or near) the head. This RF is the same as the precession frequency of the imaged nuclei at a given strength of B_0.[4] The RF pulse rotates the precessing nuclei away from their axes, thus allowing one to measure, with a receiver coil, the time it takes for the nuclei to "relax" back to their original position pointing along B_0. The spatial origin of the signal is determined using subtle, position-related changes in B_0 induced by gradient coils.[5]

7.2.1 MR Contrast in Images of Brain Structure

As shown in Table 7.1, the most common MR sequences used to characterize brain structure are T1- and T2-weighted images, DTI and MTR. We will review these in turn (for a more detailed overview of MR sequences, see Roberts and Mikulis 2007).

Contrast in structural **T1- and T2-weighted MR** images is based on local differences in proton density[6] and on the following two relaxation times: (1) longitudinal relaxation time (T1) and (2) transverse relaxation time (T2) (Fig. 7.4).

Thus, longitudinal (T1) relaxation time represents an exponential recovery of the total magnetization over time. In a complex manner, it depends on the local structural pattern (lattice)[7] assumed by the hydrogen nuclei; in the brain, the more structured the tissue, the shorter T1. Transverse (T2) reflects the local "dephasing" rate for precessing nuclei[8] due to local magnetic-field non-homogeneities, with a concomitant decay of the total MR signal; in the brain, the more structured the tissues, the more rapid the dephasing and hence, the shorter the T2.

Quantitative measurements of T1 and T2 relaxation times are not taken frequently, however (see Deoni et al. 2008 for a method suitable for a rapid voxel-wise measurement of T1 and T2 times). In the majority of MR studies, some combinations of T1- and/or T2-weighted images are acquired instead (Text Box 7.1).

---

[3] 1.5 T equals 15,000 Gauss. For comparison, the Earth's magnetic field is about 0.5 Gauss.

[4] For hydrogen, this "resonant" frequency is about 64 MHz at 1.5 T.

[5] The switching of the gradient coils generates the knocking noise heard during scanning.

[6] Proton density refers to the number of hydrogen nuclei per unit of tissue volume.

[7] Lattice is an array of points repeating periodically in three dimensions.

[8] Individual hydrogen nuclei precess at slightly different rates, leading to their magnetic moments eventually pointing in different directions ("dephasing").

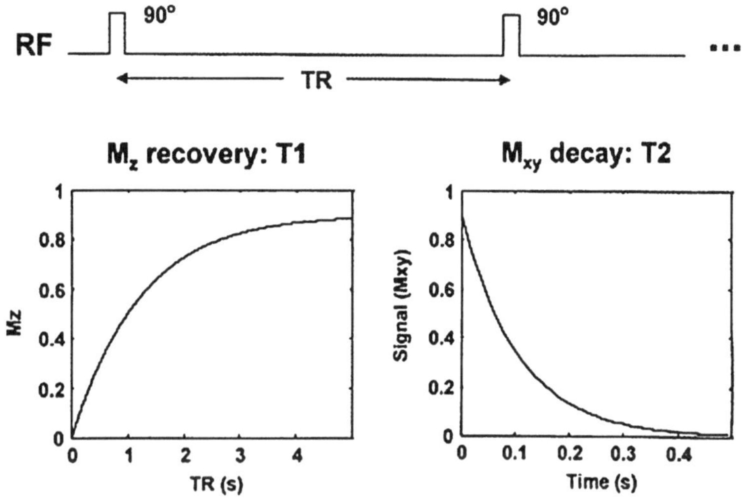

Fig. 7.4 Longitudinal (T1) and transverse (T2) relaxations times. Radio-frequency (*RF*) excitations are applied with a repetition time, with each excitation producing a magnetic resonance signal (Mxy; precession) that decays exponentially with a time constant T2 (*right panel*). Between excitations, the magnetization (Mz) also returns to its equilibrium state, pointing along the direction of the main magnetic field, with a time constant T1 (*left panel*). By manipulating the TR (repetition time) and TE (time-to-echo), the relative contributions of T1 and T2 can be manipulated to produce T1- and T2-weighted images. If the effects of both T1 and T2 are minimized (long TR and short TE), the remaining contrast is simply the proton density weighted. From Paus et al. (2001)

Text Box 7.1. Acquisition of T1- and T2-weighted images

In these acquisitions, the MR signal is repeatedly measured (with a repetition time, TR) at one time point (time-to-echo, TE) after the application of each RF pulse. Local differences in relaxation times are reflected in the image contrast simply because, at a given TR/TE combination, the MR signal has already recovered (T1) or decayed (T2) to a greater extent in regions with short T1 (or T2) and vice versa. For this reason, brain tissue with a short T1 white matter (WM) appears bright on T1-weighted images; the signal has "already recovered" (Fig. 7.4, left). In contrast, tissue with a long T2 (GM) is bright on T2-weighted images; the signal has "not yet decayed" (Fig. 7.4, right).

While proton density reflects the amount of signal-emitting nuclei (i.e. protons) present in the tissue, relaxation times, and therefore tissue "brightness" on T1- and T2-weighted images, depend on a variety of biological and structural properties of the brain tissue. Water content is one of the most important influences on T1 in the brain: the more water there is in a given tissue compartment, the longer T1—and

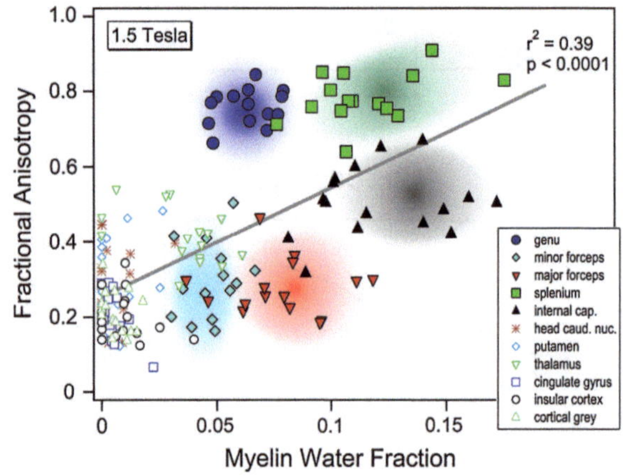

Fig. 7.5 Using the short T2-component signal to estimate myelin-water fraction, Mädler et al. (2008) found a significant correlation between this measure of myelin content and FA across different brain regions within the same individual (1.5T: $r^2 = 0.39$) but not *across individuals* within the same region. From Mädler et al. (2008)

the lower the signal on a typical T1-weighted image—will be in that compartment. In the adult brain, T1 is the longest in cerebrospinal fluid (CSF), intermediate in GM and the shortest in WM. Lipid content in WM may also influence T1, through magnetic interactions with hydrogen nuclei of the lipids, which are hydrophobic. Iron content is another important influence, primarily by changing local magnetic non-homogeneities: the higher the iron content, the shorter the T2. Finally, the anatomical arrangement of axons may influence the amount of interstitial water and, in turn, T1 values; more tightly bundled axons would have shorter T1 and therefore appear brighter on T1-weighted images.

The introduction of **diffusion tensor imaging** (DTI) in the mid-1990s opened up new avenues for in vivo studies of white-matter microstructure (Le Bihan and Basser 1995). This imaging technique allows one to estimate several parameters of water diffusion, such as mean diffusivity (MD) and fractional anisotropy (FA), in the live brain. The latter parameter reflects the degree of water-diffusion directionality; voxels containing water that moves predominantly along a single direction have higher FA. In WM, FA is believed to depend on the microstructural features of fibre tracts, including the relative alignment of individual axons, how tightly they are packed (which affects the amount of interstitial water; see above), myelin content and axon calibre. Note that the hypothetical differences in myelination are perhaps too hastily considered as an explanation for (group, age) differences in FA, to the exclusion of other possible factors (see Fig. 7.5). In most DTI studies, the signal is coming from the movement of water in the extracellular space, complicating the interpretations of the underlying neurobiology (reviewed in Paus 2010). An interesting alternative is to measure intracellular movement of a metabolite, such as *N*-acethyl-aspartate, inside the axon. This can be achieved with so-called diffusion tensor spectroscopy (DTS; Upadhyay et al. 2008). Due to the high time requirements of DTS, however, this technique is not suitable—at present—for large population-based imaging studies.

Finally, **magnetization transfer (MT) imaging** is another MR technique employed in studies of the structural properties of WM.[9] Contrast in MT images reflects the interaction between free water and water bound to macromolecules (McGowan 1999). The macromolecules of myelin are the dominant source of the MT signal in WM (Kucharczyk et al. 1994). Post-mortem studies revealed a significant positive correlation between myelin content and MTR (Schmierer et al. 2004, 2008). The fastest way to acquire MT data is to use a dual acquisition, with and without a MT saturation pulse. MTR images are calculated as the percent signal change between the two acquisitions (Pike 1996). Mean MTR values, subsequently, can be summed across all white-matter voxels constituting, for example, the four lobes of the brain.

7.2.2 MR Contrast in Images of Brain Activity

For imaging brain "activity,"[10] the most common MR parameter to measure is the blood oxygenation-level-dependent (BOLD) signal detected on T_2*-weighted MR images. This signal relies on the fact that neural "activation" is associated with an oversupply of oxygenated ("red") blood to the activated brain region; consequently, small veins draining the "activated" region contain some of the "unused" oxygenated blood ("red-veins" phenomenon). In other words, the BOLD signal reflects the proportion of oxygenated ("red") and deoxygenated ("blue") blood in a given brain region at a given moment. Variations in the oxygenated-to-deoxygenated blood ratio provide an MR contrast. Because the deoxygenated blood contains deoxyhemoglobin, which is paramagnetic, it therefore increases local non-homogeneities. That, in turn, interferes with T2 relaxation ("dephasing"; see Sect. 7.2.1).

Now what do we measure with the BOLD signal? Although there is no dispute about the relationship between local hemodynamics and brain activity, there is still little agreement about the role various neural events play in driving the hemodynamic signal. At least two issues need to be considered when interpreting the functional significance of BOLD responses to a behavioural challenge: (1) the relative importance of neuronal firing versus synaptic activity occurring in the sampled tissue and (2) the relative contributions of excitatory and inhibitory neurotransmission.

[9] MT imaging is used most often in patients with neurological disorders affecting white matter, such as multiple sclerosis (Filippi and Rocca 2007).

[10] Let me explain the reason for quotation marks here. As pointed out in this section, it is unlikely that the BOLD signal reflects neural activity understood as neurons generating action potentials, that is, firing. For this reason, the term "activation" can be used only as a metaphor. For this reason, we prefer to use terms—such as "BOLD response" or "fMRI response"—that do not imply a particular neurophysiological process.

Intuitively, a brain region requires more energy and, hence, more blood flow when neurons located in that region increase their firing rate. This notion is also conceptually attractive; if true, one would be able to take findings obtained with functional imaging of the human brain and directly correlate them with those acquired through single-unit recordings in non-human primates. Several experiments suggest, however, that firing rate is not the best predictor of local changes in hemodynamics; synaptic activity might be a better one (Text Box 7.2).

Text Box 7.2. BOLD Response: Firing Neurons or Synaptic Activity?

Using simultaneous recordings of single-unit activity, field potentials and cerebral blood flow (CBF) in the rat cerebellar cortex, Mathiesen et al. (1998) demonstrated that electrical stimulation of parallel fibres inhibited spontaneous firing of Purkinje cells located in the sampled cortex while, at the same time, it *increased* CBF and field potentials at the same location. On the other hand, a strong correlation ($r = 0.985$) was observed between postsynaptic activity, indicated by the summed *field potentials* and activity-dependent increases in CBF in the cerebellar cortex. Similar findings were obtained in the monkey cerebral (visual) cortex, where the local field potentials provided a better estimate of BOLD responses to visual stimulation than did multi-unit activity. While multi-unit activity increased only briefly at the onset of stimulation, the field-potential increase was sustained throughout the presence of the stimulus (Logothetis et al. 2001; Logothetis and Wandell 2004; see Fig. 7.6). In addition, Moore and colleagues showed that even subthreshold synaptic activity in the rat primary somatosensory cortex is accompanied by significant variations in hemodynamic signal (Moore et al. 1996, 1999).

The second issue of interest, namely the relative contribution of excitatory and inhibitory neurotransmission to changes in local hemodynamics, is less clear cut—with one study suggesting that the former drives the signal (Text Box 7.3.).

Text Box 7.3. BOLD Response: Excitation or Inhibition?

Blockage of GABAergic transmission during electrical stimulation of parallel fibres does not appear to attenuate stimulation-induced increase in local blood flow, suggesting that the increase was primarily due to the excitatory component of the postsynaptic input (Mathiesen et al. 1998). As suggested previously (e.g. Akgören et al. 1996), a direct link between excitatory neurotransmission and local blood flow may be related to the role of nitric oxide (NO) in coupling blood flow to synaptic activity. NO is one of the signals leading to dilation of small vessels in the vicinity of "active" synapses; NO is synthesized by an enzyme called NO synthase (Northington et al. 1992; Iadecola 1993). It is known that glutamate activates NO synthase through an increase in the intracellular level of calcium and that, under physiological conditions, entry of calcium into a cell is almost exclusively linked to excitatory neurotransmission.

Overall, it is likely that changes in local hemodynamics reflect a sum of excitatory postsynaptic inputs in the sample of scanned tissue. The firing rate of "output" neurons may be related to local blood flow (Heeger et al. 2000; Rees et al. 2000), but only inasmuch as it is linked in a linear fashion to excitatory postsynaptic input. Inhibitory neurotransmission may lead to decreases in CBF, indirectly, through its presynaptic effects on postsynaptic excitation.

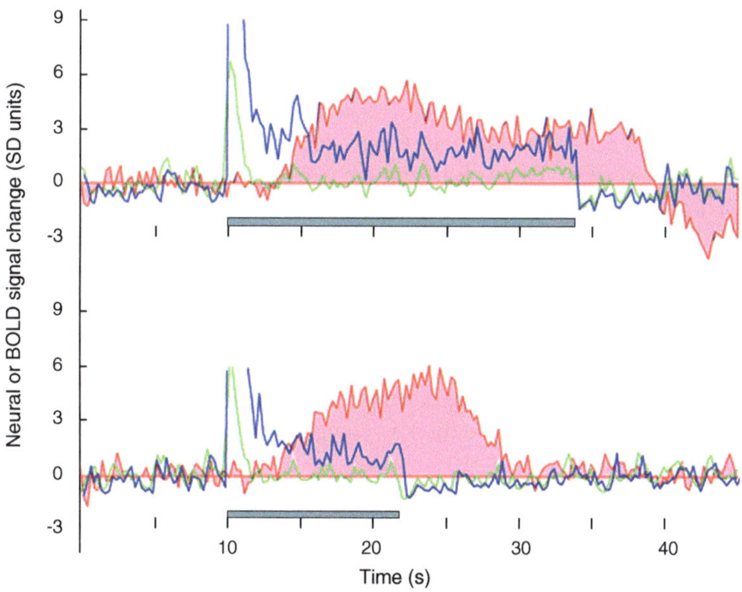

Fig. 7.6 The relationship between the BOLD response (to visual stimulation) and the local field potentials (*LFP*) and multi-unit activity (*MUA*). Responses to a 24 s (*upper*) and 12 s (*lower*) stimulation are shown. The *blue* trace measures the LFP, the *green* trace measures the MUA and the *red*-shaded trace measures the BOLD response. The MUA signal is brief and approximately the same in both experiments. The LFP and BOLD responses co-vary with the stimulus duration. At this cortical site and in 25 % of the measurements, the LFP response matched the BOLD response, but the MUA did not. There were no instances in which the MUA matched the BOLD response but the LFP did not. From Logothetis and Wandell 2004

7.3 Structural Brain Phenotypes

Here, we provide a brief overview of the key structural phenotypes that can be derived from the MR images described in Sect. 7.2.1. (see Table 7.2 for an overview) using a number of automated image-analysis algorithms (or "pipelines").

Table 7.2 Structural brain phenotypes

MRI sequence	Time (min)	Brain phenotypes
3D T1-weighted	10	Volume, thickness, folding, shape, tissue density
T2-weighted/FLAIR	4	Hyperintensities (number, volume, location)
Diffusion tensor imaging	12	Fractional anisotropy, mean diffusivity, track delineation (global, regional)
Magnetization transfer	8	Myelination index (global, regional)

Traditionally, T1-weighted images have been the richest and most flexible source of data for morphometric analyses. Over the past 20 years, a number of computational pipelines have been developed by the neuroimaging community, including the Minctools from the Montreal Neurological Institute (http://www.bic.mni.mcgill.ca/ServicesSoftware/Home Page), FMRIB Software Library (FSL) from the FMRIB laboratory at the University of Oxford (http://www.fmrib.ox.ac.uk/fsl), FreeSurfer from the Martinos Centre for Biomedical Imaging (http://surfer.nmr.mgh.harvard.edu/), BrainVisa from the NeuroSpin (http://brainvisa.info), and the LONI pipeline from UCLA (http://pipeline.loni.ucla.edu/).

These pipelines enable us to extract a large number of brain features, or phenotypes (see Fig. 7.7 for an overview) in a fully automatic fashion. There are two general classes of these features: (1) Voxel- or vertex-wise[11] features derived for each X, Y and Z location (e.g. grey- and white-matter "density" maps, cortical thickness, cortical folding) and (2) Volumetric measures (e.g. volumes of GM in frontal lobes, volume of the hippocampus, area of the corpus callosum). In the following text, we will describe the key computational steps involved in deriving these various features.

7.3.1 Density of Brain Tissues

First, T1-weighted images are linearly registered, from the acquisition ("native") space to standardized stereotaxic space—such as the average ICBM/MNI-152

[11] A vertex is a corner point of a polygon; polygons constitute a representation of a (cortical) surface.

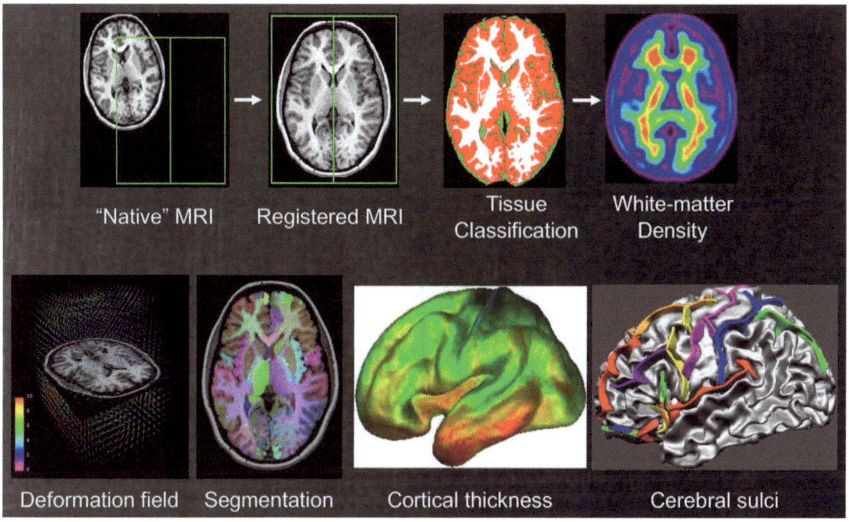

Fig. 7.7 Image-analysis pipeline. From Paus (2005)

atlas (aligned with Talairach and Tournoux space; Evans et al. 1993). The next step involves tissue classification into GM, WM and CSF. A neural-network classifier is trained, using priors[12] indicating locations that are highly likely to contain the three types of tissue (Zijdenbos et al. 2002; Cocosco et al. 2003). The tissue classification step yields three sets of binary 3D images (i.e. GM, WM and CSF). Each of the binary images can be smoothed to generate probabilistic "density" images. These maps are then used in voxel-wise analyses of age- or group-related differences in GM or WM density (see Ashburner and Friston 2000, for a methodological overview).

7.3.2 Deformation Fields

A deformation field—computed while aligning (registering) a population member to the target[13]—models the extent to which the anatomy of the individual differs from that of the population. Consequently, the set of all deformation fields captures the overall population variability. By comparing inter-population to intrapopulation deformations, we can identify and quantify anatomical differences in local

[12] The "prior" is shorthand for "a prior probability distribution". In this context, the priors bring existing knowledge to assist a particular classification algorithm.

[13] The target is usually a population average generated in a standardized (atlas) space.

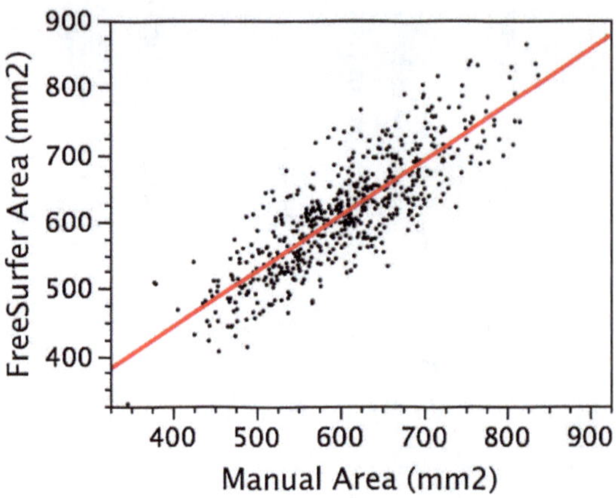

Fig. 7.8 Validation of the automatic (FreeSurfer) segmentation of the corpus callosum using manual expert-driven segmentation. From Nawaz-Khan et al. (unpublished observation)

shape between populations.[14] Unlike standard volumetric approaches, this deformation-based technique can detect differences in the morphology of brain regions that do not necessarily coincide with manually defined structures. It is therefore free from a priori considerations about anatomical nomenclature and arbitrary granularity of an atlas. In particular, it can detect changes at the substructure level or across groups of structures.

Other voxel-wise morphological information—namely differences in the local volume—can be estimated from the Jacobian[15] of the deformation field computed between the population members and the population average (Thompson et al. 2000; Chung et al. 2001).[16]

7.3.3 Regional Volumes of Brain Tissues

In order to estimate volumes of different brain regions (e.g. frontal lobes and the hippocampus), we can take advantage of the nonlinear alignment (registration) of the individual's T1-weighted images with a labelled template brain; the template contains information about the location of distinct brain regions defined a priori by an expert or generated as region-specific probabilistic maps. Then, we can compute regional volumes in an automatic fashion. Information about anatomical boundaries from the template brain can be back-projected onto each participant's

[14] The standard Hotelling T2 test is used to detect and characterize such differences (Davatzikos et al. 1996; Thompson and Toga 1997).

[15] The Jacobian is the determinant of the Jacobian matrix, which characterizes spatial relationships between vectors in (the three-dimensional) Euclidean space.

[16] Typically, one uses a standard two-sample *t* test of the log-Jacobian values to identify areas of significant local difference in volume between two populations.

brain, in native space, and intersected with the tissue classification map. In this way, we can start—for example—by counting the number of grey, white and CSF voxels per lobe (Collins et al. 1994, 1995). Or we can measure the volume of specific brain regions, such as the hippocampus and amygdala (e.g. Chupin et al. 2007; Toro et al. 2009) or the corpus callosum (Nawaz-Khan et al. unpublished observation). Note that it is important to validate automatic segmentations of brain regions against the manual expert-driven segmentation (see Fig. 7.8).

7.3.4 Cortical Thickness and Cortical Folding

FreeSurfer is the most common suite of automated tools used for reconstruction of the brain's cortical surface (Fischl and Dale 2000). FreeSurfer segments the cerebral cortex and the WM and then computes triangular meshes that recover the geometry and the topology of the pial surface and the grey/white interface of the left and right hemispheres. The way the local cortical thickness is measured is based on the difference between the position of equivalent vertices in the pial and grey/white surfaces. A correspondence between the cortical surfaces across individuals is established using a nonlinear alignment of the principal folds (sulci) in each individual's brain with those in the average brain (Fischl et al. 1999).

Approximately two-thirds of the cerebral cortex are buried in folds. To estimate the degree of cortical folding, one can simply compare the total area of the cortical

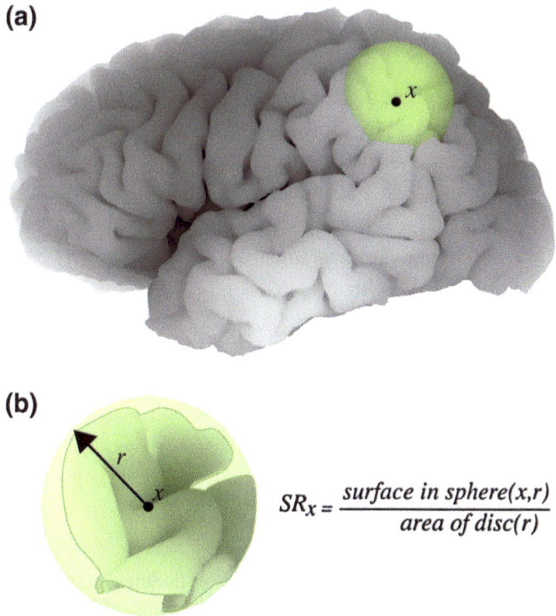

Fig. 7.9 Estimating the surface ratio. From Toro et al. (2008a)

surface to the area of its external surface. A more regional approach has been to compute the ratio between the pial contour and the outer contour in successive coronal sections (Zilles et al. 1988), which allows one to study rostro-caudal variations in folding. We extended these ideas to obtain a local (vertex-based) estimate of the degree of cortical folding; for every point x on the cortical surface, we measure the area contained in a small sphere centred at x (see Fig. 7.9; Toro et al. 2008a). If the brain were lissencephalic, the area inside the sphere would be approximately that of the disc. We estimated the local degree of folding through the surface ratio. The sphere has to be sufficiently large to encompass a few folds but small enough to make the approximation of the lissencephalic area reasonable.[17]

The measures of cortical thickness and surface ratio can be compared vertex by vertex (i.e. in the same way voxel-wise analyses of GM/WM densities and deformation fields are carried out; see Sects. 7.3.1 and 7.3.2). Alternatively, global and regional measures can be obtained by calculating mean values across all vertices contained within a particular region of interest (ROI); various anatomical parcellation schemes are available in FreeSurfer, from individual lobes to specific cortical regions (e.g. Destrieux et al. 2010).

7.3.5 *Diffusion Tensor Imaging*

Values of FA can be calculated using various tools, such as FDT (http://www.fmrib.ox.ac.uk/fsl) or ExploreDTI (http://www.exploredti.com). Diffusion-weighted images are registered with T1-weighted images and, in turn, with an atlas so that FA/MD values can be calculated for various anatomically defined compartments. An alternative approach, a so-called tract-based spatial statistics (TBSS; Smith et al. 2006), is used to test for local variations in FA/MD in a voxel-wise fashion. Here, individual FA maps are aligned nonlinearly using a method based on free-form deformations and B-splines (Rueckert et al. 1999). The cross-participant mean FA image is calculated and used to generate a white-matter tract "skeleton." Individual participants' FA values are warped onto this group skeleton for statistical comparisons by searching perpendicular from the skeleton for maximum FA values.

[17] Typically, we use r = 20 mm but our findings remain the same for r = 15 mm or r = 25 mm (Toro et al. 2008a).

7.3.6 *Magnetization Transfer Ratio*

As mentioned above (Sect. 7.2.1, last paragraph), magnetization transfer ratio (MTR) images are calculated as the percent signal change between the two acquisitions (Pike 1996). The MTR image is registered with the participant's T1-weighted image and transferred to the standardized stereotaxic space, using the same nonlinear transformation employed to align the T1-weighted image with the (ICBM-152) template. Next, we can calculate the mean MTR values for WM tissue constituting each lobe by merging the WM tissue map with the atlas-based lobar boundaries in native space. For each individual, the mean MTR values are calculated, in native space, across all WM voxels constituting a given lobar volume of WM in that participant.

7.3.7 *Multi-Modal Analysis*

Maps of MTR and FA values allow us to characterize different properties of WM. By registering these maps with the native T1-weighted image nonlinearly, we can combine information from both modalities. This enables us, for example, to measure the mean MTR and FA values of WM per lobe or fibre tract.

7.4 Functional Brain Phenotypes

As explained in Sect. 7.2.2, the most common MR signal used as an indirect index of brain "activity" is the BOLD contrast, arising from the disproportionate increase in blood flow into the region engaged functionally at a given moment (see Text Box 7.2 and 7.3 for the discussion of the possible neurophysiological correlates of this signal). In most fMRI studies, the degree of engagement of a particular brain region is measured simply by tracking the BOLD signal in time and calculating the difference between the time segments that differ from each other with regard to the presence or absence of a particular stimulus or—in more general terms—a difference in a particular well-defined behavioural state. But before we review some of the approaches used to quantify functional brain phenotypes with fMRI, let us discuss two important issues: (1) brain coverage and (2) repeatability.

First, the issue of brain **coverage**. Although we scan the entire brain, all paradigm-based fMRI studies suffer from one important shortcoming: we are able to assess functional phenotype only in brain regions that are engaged ("activated") by a given paradigm. Investigators remain blind to the functional properties of the "unstimulated" brain. For example, if the probe/paradigm involves visual stimulation (and no motor response), only brain regions engaged directly by the stimulus

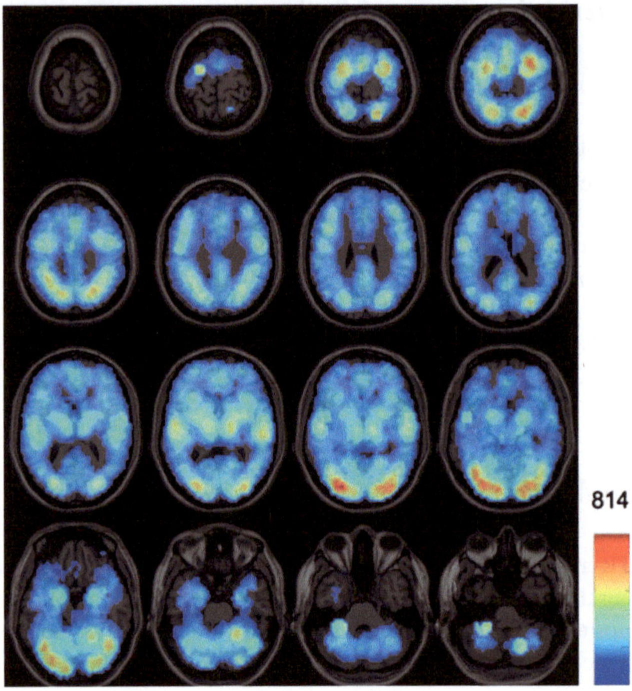

Fig. 7.10 Spatial distribution of "activations" yielded by over 3,000 functional-imaging experiments included in the BrainMap database. The colour scale indicates the number of "activations" at a given location. From Toro et al. (2008b)

(the "visual brain") can be assessed without ambiguity. Thus, in this situation, we cannot make any inferences about the functional phenotype of the "motor brain." In theory, an even coverage of the entire brain with the paradigm-based fMRI would require thousands of experiments designed to engage all possible functional circuits. Using the BrainMap database, we showed that even 3,000+ neuroimaging experiments do not cover the entire brain evenly (Toro et al. 2008b; see Fig. 7.10 for the spatial distribution of "activations" across these experiments and Fig. 7.11 for the variety of behavioural paradigms yielding these "activations").

The second issue is that of **repeatability**. As explained in the Preface, population neuroscience is interested in uncovering factors that shape the brain, both from within (genes) and without (social and physical environment). This puts one key requirement on the phenotype: the presence of high test–retest reliability when measured across time (days, weeks) in the same individual. In other words, we need a phenotype that is repeatable (stable over a period of time) so that we can identify with some confidence the genes and/or long-acting environmental/experiential factors that shaped it.

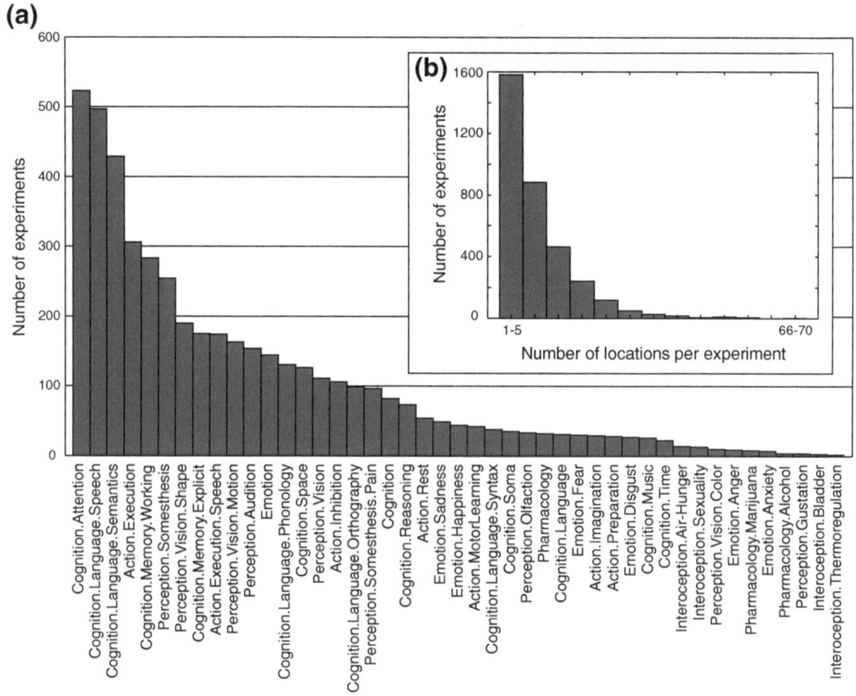

Fig. 7.11 Distribution of the different cognitive domains represented by the experiments after the BrainMap classification (**a**). Histogram of the number of "activated" locations per experiment (**b**). On average, experiments reported eight locations; a decreasing number of experiments reported large numbers of locations. From Toro et al. (2008b)

Let us now examine the repeatability of fMRI, as measured in the same individual.[18] As reviewed thoroughly by Caceres et al. (2009), we can assess repeatability of fMRI measurements by calculating intraclass correlation coefficients (ICCs).[19] This is done by comparing differences in between-subject and within-subject (e.g. across two sessions) variability, relative to the total variation in the sample (the sum of within- and between-subject variability). In fMRI studies, there are several ways one can calculate ICCs for the BOLD response[20]: (1) for all voxels across the entire brain; (2) for a subset of voxels, constituting a given ROI; and (3) for the mean (or median) BOLD response in a given ROI (Caceres et al. 2009). Evaluating the median of the ICCs (medICCs) calculated for all voxels constituting a given ROI may be particularly useful (Caceres et al. 2009). Using fMRI data collected with two different paradigms in 10 participants

[18] Note that we are not interested here in the repeatability of average (group) "activation" maps.

[19] The most common type of ICC used in this context is the third ICC (Shrout and Fleiss 1979).

[20] The magnitude of the BOLD response is typically assessed in a particular contrast (difference) between two conditions.

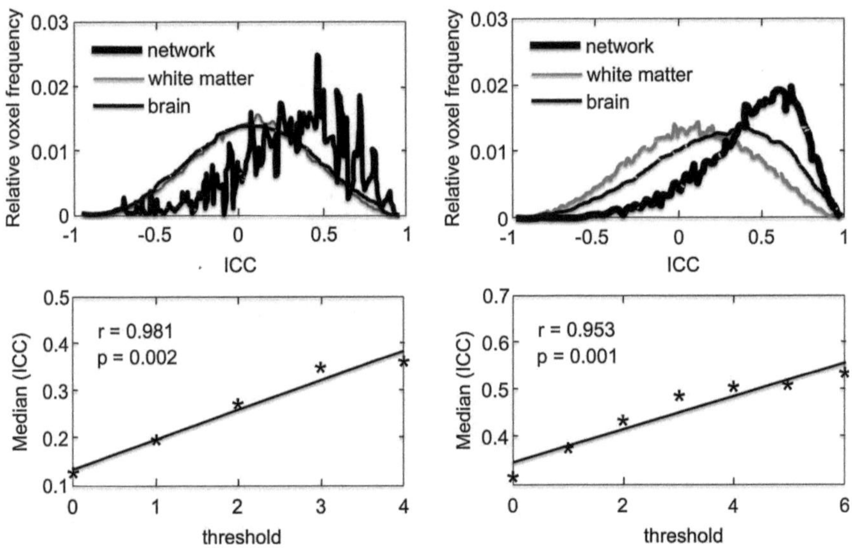

Fig. 7.12 Intraclass correlation coefficients (*ICCs*) calculated from fMRI data obtained in 10 individuals tested on two occasions (3 months apart), using two different paradigms: an auditory task (*left panel*) and an N-back task (*right panel*). In the *bottom row*, the median of the "network" distribution is plotted against the threshold that defines the region. A strong correlation is shown for both tasks. The distribution, corresponding to the white matter with negligible median, is also shown for both tasks. From Caceres et al. (2009)

across two sessions (three months apart), Caceres et al. (2009) showed that medICCs were ~0.5 for both tasks in the "activated" ROIs; medICCs correlated strongly with the *t* threshold in the "activated" ROIs and were centred around zero in the WM (Fig. 7.12). It is also important to note that the medICCs varied widely across different ROIs (Fig. 7.13) and, in this sample of 10 participants, their 95 % confidence intervals were quite wide.

Using a similar approach, a number of other studies evaluated ICCs in different behavioural paradigms involving, for example, reward and face processing (Plichta et al. 2012) or task switching (Koolschijn et al. 2011).

The test–retest reliability of the BOLD response across these studies has been characterized as "fair" in adults and adolescents, and "poor" in children.[21] These values contrast with those reported by Aron et al. (2006), who evaluated ICCs in eight participants scanned twice (a year apart) during probabilistic classification learning; they reported voxel-based ICCs in the "excellent" range (~0.8). But as can be seen in Fig. 7.14., relatively high ICC values were also found in the "non-activated" regions of the brain, raising an interesting issue of factors—possibly independent of task-related neural response—that underlie such session-to-session, within-subject consistency in brain regions not engaged by the task.

[21] Test–retest reliability: poor—ICCs <0.4; fair—0.4–0.75; excellent—>0.75 (Fleiss 1986).

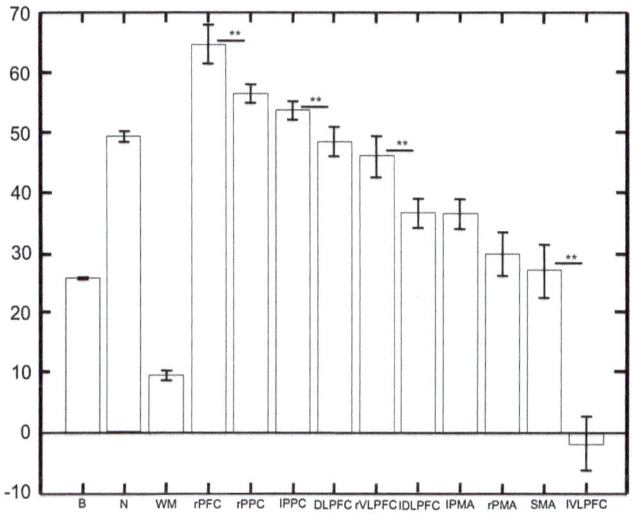

Fig. 7.13 Medians of the distributions of intraclass correlation coefficients in each (ROI; rFPC, rPPC, lPPC, rDLPFC, rVLPFC, lDLPFC, lPMA, SMA, lVLPFC) for the N-back task (same experiment as shown in Fig. 7.12). The figure also shows the reliabilities for the brain (*B*), network (*N*) and white matter (*WM*). A multiple comparison test was run across all the ROIs ($\alpha = 0.01/45$). Significant differences are shown. From Caceres et al. (2009)

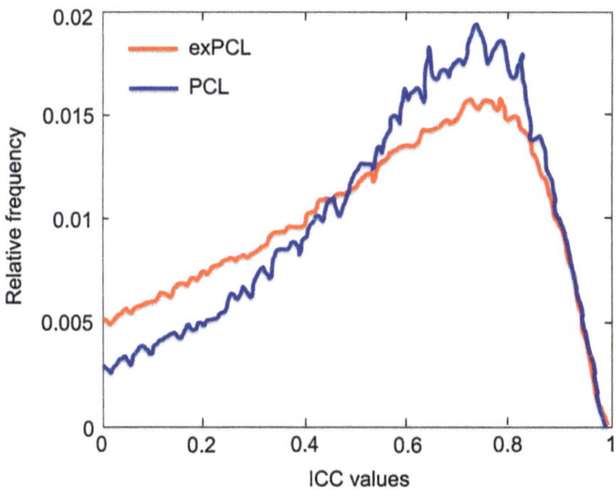

Fig. 7.14 Relative frequency histogram of intraclass correlation coefficients (*ICC*) for the probabilistic classification learning (*PCL*) network and areas outside the PCL network (exPCL); zero values and negative values are excluded in both cases. Values within the PCL network are significantly higher than for those in ex-PCL (chi-square test for the difference between distributions, $P < 0.0001$), confirming that test–retest reliability is greater in areas engaged by the task. From Aron et al. (2006)

Others have observed a similar phenomenon. For example, Caceres et al. (2009) noted regions of high ICCs that did *not* overlap with regions of consistent BOLD response, as defined by the group t maps; they speculated that these regions might show a nonlinear response to the paradigm (Caceres et al. 2009). Raemaekers et al. (2012) proposed a new modelling-based approach for isolating global variations in the BOLD response, which may "scale the amplitudes of BOLD responses (and their estimates) with roughly the same factor throughout the brain." They scanned 21 participants twice (one week apart) in two different paradigms and observed significant differences between the two sessions in the BOLD response across the brain (14–23 %), which they attributed to "global effects" (Raemaekers et al. 2012). They concluded that "this (global) variability needs to be taken into account in studies correlating individual differences in amplitudes in BOLD responses with … behavioural, personality traits, or illness measured across subjects" (Raemaekers et al. 2012).

Altogether the above studies suggest that while the test–retest reliability of the amplitude of a task-related BOLD response is "fair" at best, it is lower in regions with weak "activation" (revealed by group maps). The test–retest reliability might be influenced by individual variations in "global" scaling of the BOLD response due to both technical (scanner instabilities, head motion) and biological (arousal, habituation) factors. Addressing these issues would undoubtedly increase the robustness of using the BOLD response as a phenotype.

7.5 Assessment of Cognitive Abilities and Mental Health

Measuring a brain response to a particular stimulus or cognitive challenge in the scanner is not the only way to link brain and behaviour. As explained briefly in Sect. 7.1, an unlimited array of structure–function relationships can be explored by evaluating associations between various structural phenotypes and behaviours assessed *outside* the scanner. Furthermore, repeating such measurements over time—or before and after a particular intervention—will create opportunities for testing directionality and/or causality in such structure–function relationships (see Chap. 9 for more details).

Clearly, cognitive abilities and mental health can be assessed in many ways; there is no shortage of instruments available for this purpose. Key issues to be considered when putting together a phenotyping protocol are as follows: (1) breadth of the assessment and (2) participant's time.

Typically, population-based studies do not target one specific cognitive domain (e.g. attention) or mental-health problem (e.g. depression). In fact, given the richness of molecular and brain phenotypes, and of inter-dependencies that exist across different cognitive and mental-health domains, it would be counterproductive to restrict a study to one area. This is why it is desirable to "cast a wide net," so the individual's cognitive abilities and mental health are characterized as broadly as possible within a reasonable time frame.

7.5.1 Assessment of Cognitive Abilities

The assessment of cognitive abilities can be carried out by a trained research assistant in a face-to-face examination or in a fully automated (i.e. web-based) manner, executed either in the laboratory or at home.

The most common tool for evaluating general cognitive abilities in a **face-to-face** manner is an assessment of general intelligence using, for example, a version of the Wechsler Intelligence Scale (e.g. WAIS-IV for adults or WISC-IV for children). The Wechsler intelligence tests take about 1 h to administer and provide standardized scores that characterize various cognitive abilities across a number of domains, including verbal comprehension, perceptual reasoning, processing speed and working memory. Tests of other domains, such as short-term and long-term memory, math abilities, cognitive control ("executive" functions) and motor skills, are often added to the assessment of general intelligence. In the Saguenay Youth Study, we have devoted a total of 6 h (i.e. one school day) to the cognitive assessment of each of the 1,000 adolescent participants (Table 7.3).

Computer-based tests of cognition are increasingly being used in clinical trials, as well as in large-scale, population-based cognition studies, with the expectation that they may be more sensitive and administered more efficiently than traditional neuropsychological tests. In our studies, we use a web-based platform for the assessment of cognitive abilities developed by Adam Hampshire and Adrian Owen: the Cambridge Brain Sciences test (cambridgebrainscience.com). This platform, which can be used either in the laboratory or at home (over the Internet), currently consists of 12 computer-based tests of executive function, memory, learning and attention, which together take approximately 35 min to complete (see Table 7.4). All of these tests draw heavily on well-established neuropsychological tools, although, unlike traditional cognitive batteries, the computerized tests generate novel stimuli instantaneously and adapt dynamically to accommodate any ability level. Reliable indices of performance can typically be acquired in 2–3 min per test using the alternating staircase method. The tests are also designed to be operator-free. Pilot studies reveal high correlations between scores obtained in both young and old individuals, when tested alone at home and when supervised in laboratory conditions. The tests have been validated in young controls and patients with Parkinson's disease. Population norms are available from two large-scale public trials involving more than 100,000 participants (Hampshire et al. 2011; Owen et al. 2012).

A similar (adaptive) approach has been applied in the construction of measures suitable for characterizing behaviour across a wide age-range (3 to 85 years) in four domains, namely motor behaviour, sensation, cognition and emotion; each domain-based battery takes about 30 minutes to complete (http://www.nihtoolbox.org).

Table 7.3 Face-to-face assessment of cognitive abilities in adolescents employed in the Saguenay youth study (from Pausova et al. 2007)

Test	Time of day
MNI: Handedness	9.00
WISC: Picture completion	9.03
WISC: Information	9.08
CMS: Stories immediate	9.15
CMS: Dot locations immediate	9.19
WISC: Coding (B)	9.26
WISC: Symbol search	9.29
Ruff 2–7	9.33
Digit span	9.41
Grooved pegboard	9.46
CMS: Stories recall	9.55
CMS: Stories recognition	9.59
CMS: Dot locations recall	10.00
WASI: Similarities	10.11
WASI: Picture arrangement/matrices	10.19
WISC: Arithmetic	10.31
WASI: Block design	10.36
WASI: Vocabulary	10.46
WISC: Object assembly	10.52
WISC: Comprehension	11.01
Stroop	11.04
Word fluency	11.07
W-J: Math	11.12
W-J: Math fluency	11.22
W-J: Reading comprehension	11.25
W-J: Spelling	11.45
	Lunch
Number sense	1.00
Frequency modulation threshold	1.10
Phonological learning	1.20
Body size perception (self)	1.50
Body size perception (discrimination)	2.03
Tapping	2.31
Bimanual coordination	2.36
	Break
Action observation	3.25
Self-ordered pointing	3.30
Newman's card playing	3.50
Pollak faces	4.00
Anagram task	4.20

MNI Montreal neurological institute; *WISC* wechsler intelligence scale for children; *WASI* wechsler abbreviated scale of intelligence; *CMS* children's memory scale; *W-J*, Woodcock-Johnson

Table 7.4 Cognitive assessment: the web-based Cambridge brain sciences battery

Instrument	Time (min)	Process/domain
Cattell's culture fair test (CCFT)	5	General intelligence (g)
Grammatical reasoning	1.5	Reasoning ability (based on the Baddeley grammatical reasoning test)
Odd one out	3	Deductive reasoning (problems from CCFT)
Digit span	2	Short-term verbal memory
Self-ordered search	3	Spatial working memory
Visuo-spatial working memory	2	Visuo-spatial working memory
Paired associate learning (PAL)	2	Verbal memory
Spatial planning (a variant of tower of London)	3	Planning
Feature match	1.5	Attention
Colour word remapping (a variant of Stroop)	1.5	Resistance to interference
Risk tolerance (a variant of balloon analogue risk task)	2	Risk/impulsivity
Reward/punishment (a variant of Iowa gambling task)	4	Reward/punishment sensitivity

7.5.2 Assessment of Mental Health

Assessment of mental health can be carried out by a trained professional in the form of a structured psychiatric interview or through questionnaires designed to elicit self-reports about the participant behaviour and mental-health problems.

Structured psychiatric interviews have the advantage of allowing investigators to determine a tentative clinical diagnosis consistent with the current diagnostic criteria (e.g. DSM-IV). This, of course, is essential in studies that specifically recruit individuals with particular psychiatric disorders and their matched controls (i.e. case–control studies). In community-based studies of the general population, the use of psychiatric interviews is helpful, especially in the context of establishing a tentative clinical diagnosis of common disorders, such as mood and substance-use disorders. It is also helpful when planning a long-term follow-up of all participants so that one can identify temporal relationships between exposures (e.g. a stressful life event), mediators (e.g. volume of the anterior cingulate cortex) and outcomes (e.g. presence/absence of clinical depression), and their directionality. There are various validated tools for conducting structured psychiatric interviews suitable for adults (Table 7.5).

In our work, we use the Mini-international neuropsychiatric interview (MINI Plus); it is a validated structured psychiatric interview for current and lifetime DSM-IV and ICD-10 psychiatric disorders (Sheehan et al. 1998). In our experience, this interview takes between 20 and 60 min when administered in middle-aged adults recruited from a general population. The shorter version of this instrument (MINI) focuses only on the current diagnoses and takes about 15 min

Table 7.5 Structured psychiatric interviews for adults

Interview	Duration (min)	Time frame	Diagnostic output
PSE	15–60	Last month	Descriptive syndromes
DIS	45–75	Lifetime; past month; past 6 months; past year	DSM-III-R diagnoses
SADS	90–120	Previous weeks	RDC categories
SCID	45–60	Current episode	DSM-III-R diagnoses
CIDI	120–180	Current and lifetime	ICD-10 and DSM-III-R
MINI	15	Current; a few lifetime	DSM-IV and ICD-10
MINI Plus	45–60	Current and lifetime	DSM-IV and ICD-10
MINI Screen	5	Current	DSM-IV and ICD-10 Screen
SDDS	3–10	Current	DSM-III-R diagnoses
PRIME-MD	8	Current	DSM-III-R diagnoses

PSE present state examination; *DIS* diagnostic interview schedule; *SADS* schedule for affective disorders; *SCID* structured clinical interview for DSM-III-R diagnoses; *CIDI* composite international diagnostic interview for ICD-10; *PRIME-MD* primary care evaluation of mental disorders; *SDDS* symptom-driven diagnostic system; *MINI* mini-international neuropsychiatric interview; *RDC* research diagnostic criteria. Adapted from Sheehan et al. (1998)

Table 7.6 Diagnostic outputs provided by mini-international neuropsychiatric interview (*MINI*)

Disorder diagnoses available on the MINI[a]

Disorder	Time frame
Major depressive disorder	Past 2 weeks
Dysthymic disorder	Past 2 years
Suicidality	Current
Mania	Lifetime and current
Panic disorder	Lifetime and current
Agoraphobia	Current
Social phobia	Current
Specific phobia	Current
Obsessive–compulsive disorder	Current
Generalized anxiety disorder	Current
Alcohol dependence	Current
Alcohol abuse	Current
Drug dependence (non-alcohol)	Current
Drug abuse (non-alcohol)	Current
Psychotic disorder	Lifetime and current
Anorexia nervosa	Past 3 months
Bulimia	Past 3 months
Post-traumatic stress disorder	Current
Antisocial personality disorder	Lifetime

[a] Current is defined as "in the past month" for all diagnoses except for generalized anxiety disorder (past 6 months) and alcohol and drug) abuse/dependence (12 months). From Sheehan et al. (1998)

Table 7.7 Lifetime prevalence of commonly occurring mental disorders (in United States)

	Total		18–29 years		30–44 years		45–59 years		60+ years	
	%	(se)	%	(se)	%	(se)	%	(se)	%	(se)
Panic disorder	4.7	(0.2)	4.4	(0.4)	5.7	(0.5)	5.9	(0.4)	2.0	(0.4)
Agoraphobia without panic	1.4	(0.1)	1.1	(0.2)	1.7	(0.3)	1.6	(0.3)	1.0	(0.3)
Specific phobia	12.5	(0.4)	13.3	(0.8)	13.9	(0.8)	14.1	(1.0)	7.5	(0.7)
Social phobia	12.1	(0.4)	13.6	(0.7)	14.3	(0.8)	12.4	(0.8)	6.6	(0.5)
Generalized anxiety disorder	5.7	(0.3)	4.1	(0.4)	6.8	(0.5)	7.7	(0.7)	3.6	(0.5)
PTSD	6.8	(0.4)	6.3	(0.5)	8.2	(0.8)	9.2	(0.9)	2.5	(0.5)
OCD	1.6	(0.3)	2.0	(0.5)	2.3	(0.9)	1.3	(0.6)	0.7	(0.4)
SAD	5.2	(0.4)	5.2	(0.6)	5.1	(0.6)	–	–	–	–
Any anxiety disorder	28.8	(0.9)	30.2	(1.1)	35.1	(1.4)	30.8	(1.7)	15.3	(1.5)
Major depressive disorder	16.6	(0.5)	15.4	(0.7)	19.8	(0.9)	18.8	(1.1)	10.6	(0.8)
Dysthymia	2.5	(0.2)	1.7	(0.3)	2.9	(0.4)	3.7	(0.7)	1.3	(0.3)
Bipolar I-II disorders	3.9	(0.2)	5.9	(0.6)	4.5	(0.3)	3.5	(0.4)	1.0	(0.3)
Any mood disorder	20.8	(0.6)	21.4	(1.1)	24.6	(0.9)	22.9	(1.2)	11.9	(1.0)
ODD	8.5	(0.7)	9.5	(0.7)	7.5	(0.8)	–	–	–	–
CD	9.5	(0.8)	10.9	(0.3)	8.2	(0.8)	–	–	–	–
ADHD	8.1	(0.6)	7.8	(0.6)	8.3	(0.9)	–	–	–	–
Intermittent explosive disorder	5.2	(0.3)	7.4	(0.9)	5.7	(0.6)	4.9	(0.4)	1.9	(0.5)
Any impulse control disorder	25.8	(1.1)	26.8	(0.9)	23.0	(1.3)	–	–	–	–
Alcohol abuse	13.2	(0.6)	14.3	(1.0)	16.3	(1.1)	14.0	(1.1)	6.2	(0.7)
Alcohol dependence	5.4	(0.3)	6.3	(0.7)	6.4	(0.6)	6.0	(0.7)	2.2	(0.4)
Drug abuse	7.9	(0.4)	10.9	(0.9)	11.9	(1.0)	6.5	(0.6)	0.3	(0.2)
Drug dependence	3.0	(0.2)	3.9	(0.5)	4.9	(0.6)	2.3	(0.4)	0.2	(0.1)
Any substance disorder	14.6	(0.6)	16.7	(1.1)	18.0	(1.1)	15.3	(1.0)	6.3	(0.7)
Any	46.4	(1.1)	52.4	(1.7)	55.0	(1.6)	46.5	(1.8)	26.1	(1.7)
Two or more disorders	27.7	(0.9)	33.9	(1.3)	34.0	(1.5)	27.0	(1.6)	11.6	(1.0)
Three or more disorders	17.3	(0.7)	22.3	(1.2)	22.5	(1.1)	15.9	(1.3)	5.3	(0.7)

ADHD attention-deficit/hyperactivity disorder; *CD* conduct disorder; *OCD* obsessive–compulsive disorder; *ODD* oppositional-defiant disorder; *PTSD* post-traumatic stress disorder; *SAD* separation anxiety disorder. From Kessler and Wang (2008)

to administer (Sheehan et al. 1998). In children and adolescents, we use MINI-Kid (Sheehan et al. 2010). Overall, the family of the MINI instruments provides a broad diagnostic output relevant for most of the common mental-health disorders (Sheehan et al. 1998; Table 7.6).

Questionnaire-based assessments represent an excellent technique for collecting information about the mental health in population-based studies. In adolescents and adults, we typically ask participants to provide answers to a number of questions about their mental health (i.e. they self-report). In children, we often seek an evaluation of mental-health symptoms and behaviour by their parents and/or teachers. Two types of instruments are available in terms of the breath/specificity vis-à-vis mental health: (1) general and (2) targeted.

Table 7.8 Questionnaires suitable, in adolescents and middle-aged adults, for the assessment of symptoms related to anxiety, mood, impulse control and substance use

Condition	Questionnaire
Anxiety	GRIPado, GRIPadult
	NEO-PI-R: Neuroticism
Mood	Centrum for epidemiological studies—depression (CES-D)
Impulse control	Conners' adult ADHD rating scale
	Adult ADHD self-report scale (ASRS-v1.1)
	NEO-PI-R: Agreeableness
Substance use and other addictive behaviours	Fagerström's nicotine dependence scale
	Nicotine dependence syndrome scale
	Alcohol use disorder identification test
	Michigan alcoholism screening test
	Subjective response to ethanol
	Drug abuse screening test
	European school survey project on alcohol and drugs
	South Oklahoma gambling screen revised
	Internet addiction test
	Yale food addiction scale

Instruments of the *general* kind allow investigators to make an "epidemiological" diagnosis across a broad array of psychiatric disorders, again defined by clinical criteria (e.g. DSM-IV), by asking a number of questions about the key symptoms associated with each disorder. In a way, such tools are the "poor cousins" of the structured psychiatric interviews discussed above. They provide a screen for the presence of a *possible* clinical diagnosis and, at the same time, allow one to calculate the number of symptoms reported by the participant in many mental-health domains. In adults, for this purpose, we use the Adult Self-Report (Achenbach and Rescorla 2003). In adolescents, we have been using the diagnostic interview schedule for children (DISC) predictive scales (DPS) (Lucas et al. 2001).

The *targeted* instruments, on the other hand, delve deeper into the symptoms of a particular condition, such as depression or alcohol abuse, often allowing one to establish its onset, duration and severity. Although the choice of such targeted questionnaires depends to some extent on the type of the sampled population (e.g. adolescents vs. elderly) and the specificity of the primary questions asked by the investigators, it is always helpful to cover mental-health conditions that have a relatively high prevalence in the general populations (see Table 7.7).

There is a large number of instruments available for this purpose; Table 7.8 provides a list of those we use in our population-based studies carried out in adolescent and middle-aged populations.

In summary, systems-level phenomics—afforded on a large scale by the structural and functional imaging of the human brain, as well as by the face-to-face and web-based assessments of cognitive abilities and mental health—provides a wealth of "outcomes" for identifying key events in our genes and environment that shape the human brain and behaviour. A number of studies of community-based samples of the general population are under way. The next chapter will describe in some detail the design of some of these studies.

References

Achenbach T, Rescorla LA (2003) Manual for the ASEBA adult forms and profiles. University of Vermont, Research Center for Children, Youth, and Families, Burlington

Akgoren N, Dalgaard P, Lauritzen M (1996) Cerebral blood flow increases evoked by electrical stimulation of rat cerebellar cortex: relation to excitatory synaptic activity and nitric oxide synthesis. Brain Res 710 (1–2):204–214. doi:0006-8993(95)01354-7 [pii]

Anblagan D, Jones NW, Costigan C, Parker AJJ, Allcock K, Aleong R, Coyne L, Deshpande R, Raine Fenning N, Bugg G, Roberts N, Pausova Z, Paus T, Gowland PA. Maternal Smoking during Pregnancy and Fetal Organ Growth: a Magnetic Resonance Imaging Study (unpublished observation)

Ashburner J, Friston KJ (2000) Voxel-based morphometry—the methods. NeuroImage 11(6 Pt 1):805–821

Aron AR, Gluck MA, Poldrack RA (2006) Long-term test-retest reliability of functional MRI in a classification learning task. Neuroimage 29(3):1000–1006. doi:S1053-8119(05)00598-7 [pii] 10.1016/j.neuroimage.2005.08.010

Caceres A, Hall DL, Zelaya FO, Williams SC, Mehta MA (2009) Measuring fMRI reliability with the intra-class correlation coefficient. Neuroimage 45(3):758–768. doi:S1053-8119(08)01327-X [pii] 10.1016/j.neuroimage.2008.12.035

Chung MK, Worsley KJ, Paus T, Cherif C, Collins DL, Giedd JN et al (2001) A unified statistical approach to deformation-based morphometry. NeuroImage 14(3):595–606

Chupin M, Hammers A, Bardinet E, Colliot O, Liu RS, Duncan JS, Garnero L, Lemieux L (2007) Fully automatic segmentation of the hippocampus and the amygdala from MRI using hybrid prior knowledge. Med Image Comput Assist Interv 10(Pt 1):875–882

Cocosco CA, Zijdenbos AP, Evans AC (2003) A fully automatic and robust brain MRI tissue classification method. Med Image Anal 7(4):513–527

Collins DL, Neelin P, Peters TM, Evans AC (1994) Automatic 3D intersubject registration of MR volumetric data in standardized Talairach space. J Comput Assist Tomo 18(2):192–205

Collins D, Holmes C, Peters T, Evans A (1995) Automatic 3-D model-based neuroanatomical segmentation. Hum Brain Mapp 3:190–208

Davatzikos C, Vaillant M, Resnick SM, Prince JL, Letovsky S, Bryan RN (1996) A computerized approach for morphological analysis of the corpus callosum. J Comput Assist Tomo 20(1):88–97

Deoni SC, Rutt BK, Arun T, Pierpaoli C, Jones DK (2008) Gleaning multicomponent T1 and T2 information from steady-state imaging data. Magn Reson Med 60(6):1372–1387. doi:10.1002/mrm.21704

Destrieux C, Fischl B, Dale A, Halgren E (2010) Automatic parcellation of human cortical gyri and sulci using standard anatomical nomenclature. Neuroimage 53(1):1–15. doi:S1053-8119(10)00854-2 [pii] 10.1016/j.neuroimage.2010.06.010

Draganski B, Gaser C, Busch V, Schuierer G, Bogdahn U, May A (2004) Neuroplasticity: changes in grey matter induced by training. Nature 427(6972):311–312

Evans A, Collins D, Mills S (1993) 3D statistical neuroanatomical models from 305 MRI volumes. Nuclear Science Symposium and Medical Imaging Conference, San Francisco, p 1813–1817

Filippi M, Rocca MA (2007) Magnetization transfer magnetic resonance imaging of the brain, spinal cord, and optic nerve. Neurotherapeutics 4(3):401–413

Fischl B, Dale AM (2000) Measuring the thickness of the human cerebral cortex from magnetic resonance images. Proc Natl Acad Sci USA 97(20):11050–11055

Fischl B, Sereno MI, Tootell RB, Dale AM (1999) High-resolution intersubject averaging and a coordinate system for the cortical surface. Hum Brain Mapp 8(4):272–284. doi:10.1002/(SICI)1097-0193(1999)8:4<272::AID-HBM10>3.0.CO;2-4 [pii]

Fleiss JL (1986) The design and analysis of clinical experiments. Wiley, New York

Hampshire A, Highfield R, Parkin B, Owen AM (2011) Fractionating human intelligence (submitted)

Heeger DJ, Huk AC, Geisler WS, Albrecht DG (2000) Spikes versus BOLD: what does neuroimaging tell us about neuronal activity? Nat Neurosci 3(7):631–633. doi:10.1038/76572

Iadecola C (1993) Regulation of the cerebral microcirculation during neural activity: is nitric oxide the missing link? Trends Neurosci 16(6):206–214

Kessler RC, Wang PS (2008) The descriptive epidemiology of commonly occurring mental disorders in the United States. Annu Rev Public Health 29:115–129. doi:10.1146/annurev.publhealth.29.020907.090847

Koolschijn PC, Schel MA, de Rooij M, Rombouts SA, Crone EA (2011) A three-year longitudinal functional magnetic resonance imaging study of performance monitoring and test-retest reliability from childhood to early adulthood. J Neurosci 31(11):4204–4212

Kucharczyk W, Macdonald PM, Stanisz GJ, Henkelman RM (1994) Relaxivity and magnetization transfer of white matter lipids at MR imaging: importance of cerebrosides and pH. Radiology 192(2):521–529

Le Bihan D, Basser PJ (1995) Molecular diffusion and nuclear magnetic resonance. In: LeBihan D (ed) Diffusion and perfusion magnetic resonance imaging. Raven Press, New York, p 5–17

Logothetis NK, Wandell BA (2004) Interpreting the BOLD signal. Annu Rev Physiol 66:735–769. doi:10.1146/annurev.physiol.66.082602.092845

Logothetis NK, Pauls J, Augath M, Trinath T, Oeltermann A (2001) Neurophysiological investigation of the basis of the fMRI signal. Nature 412(6843):150–157. doi:10.1038/35084005 35084005 [pii]

Lucas CP, Zhang H, Fisher PW, Shaffer D, Regier DA, Narrow WE, Bourdon K, Dulcan MK, Canino G, Rubio-Stipec M, Lahey BB, Friman P (2001) The DISC predictive scales (DPS): efficiently screening for diagnoses. J Am Acad Child Adolesc Psychiatry 40(4):443–449

Madler B, Drabycz SA, Kolind SH, Whittall KP, MacKay AL (2008) Is diffusion anisotropy an accurate monitor of myelination? Correlation of multicomponent T2 relaxation and diffusion tensor anisotropy in human brain. Magn Reson Imaging 26(7):874–888. doi:10.1016/j.mri.2008.01.047 S0730-725X(08)00083-0 [pii]

Mathiesen C, Caesar K, Akgoren N, Lauritzen M (1998) Modification of activity-dependent increases of cerebral blood flow by excitatory synaptic activity and spikes in rat cerebellar cortex. J Physiol 512(Pt 2):555–566

McGowan JC (1999) The physical basis of magnetization transfer imaging. Neurology 53(5 Suppl 3):S3–7

Moore CI, Sheth, B, Basu A, Nelson S, Sur M (1996) What is the neural correlate of the optical imaging signal? Intracellular receptive field maps and optical imaging in rat barrel cortex. Soc Neurosci

Moore CI, Nelson SB, Sur M (1999) Dynamics of neuronal processing in rat somatosensory cortex. Trends Neurosci 22(11):513–520. doi:S0166-2236(99)01452-6

Nawaz-Khan I, Qiu N, Leonard G, Perron M, Pike GB, Richer L, Veillette S, Ferguson E, Pausova Z, Paus T. Validation of FreeSurfer-based volumetric estimates of the corpus callosum (unpublished observation)

Northington FJ, Matherne GP, Berne RM (1992) Competitive inhibition of nitric oxide synthase prevents the cortical hyperemia associated with peripheral nerve stimulation. Proc Natl Acad Sci U S A 89(14):6649–6652

Owen AM, Hampshire A, Grahn JA, Stenton R, Dajani S, Burns AS, Howard RJ, Ballard CG (2012) Putting brain training to the test. Nature 465(7299):775–778. doi:10.1038/nature09042 (nature09042)

Paus T (2005) Mapping brain maturation and cognitive development during adolescence. Trends Cogn Sci 9(2):60–68

Paus T (2010) Population neuroscience: why and how. Hum Brain Mapp 31(6):891–903

Paus T, Marrett S, Worsley KJ, Evans AC (1995) Extraretinal modulation of cerebral blood flow in the human visual cortex: implications for saccadic suppression. J Neurophysiol 74(5):2179–2183

Paus T, Collins DL, Evans AC, Leonard G, Pike B, Zijdenbos A (2001) Maturation of white matter in the human brain: a review of magnetic resonance studies. Brain Research Bulletin 54(3):255–266

Paus T, Bernard M, Chakravarty MM, Davey Smith G, Gillis J, Lourdusamy A, Melka MG, Leonard G, Pavlidis P, Perron M, Pike GB, Richer L, Schumann G, Timpson N, Toro R, Veillette S, Pausova Z (2012) KCTD8 gene and brain growth in adverse intrauterine environment: a genome-wide association study. Cereb Cortex 22(11):2634–2642

Pausova Z, Paus T, Abrahamowicz M, Almerigi J, Arbour N, Bernard M, Gaudet D, Hanzalek P, Hamet P, Evans AC, Kramer M, Laberge L, Leal S, Leonard G, Lerner J, Lerner RM, Mathieu J, Perron M, Pike B, Pitiot A, Richer L, Seguin JR, Syme C, Toro R, Tremblay RE, Veillette S, Watkins K (2007) Genes, maternal smoking, and the offspring brain and body during adolescence: design of the Saguenay youth study. Hum Brain Mapp 28(6):502–518

Pike GB (1996) Pulsed magnetization transfer contrast in gradient echo imaging: a two-pool analytic description of signal response. Magn Reson Med 36(1):95–103

Plichta MM, Schwarz AJ, Grimm O, Morgen K, Mier D, Haddad L, Gerdes AB, Sauer C, Tost H, Esslinger C, Colman P, Wilson F, Kirsch P, Meyer-Lindenberg A (2012) Test-retest reliability of evoked BOLD signals from a cognitive-emotive fMRI test battery. Neuroimage 60(3):1746–1758

Raemaekers M, du Plessis S, Ramsey NF, Weusten JM, Vink M (2012) Test-retest variability underlying fMRI measurements. Neuroimage 60(1):717–727. doi:10.1016/j.neuroimage.2011.11.061 S1053-8119(11)01343-7 [pii]

Rees G, Friston K, Koch C (2000) A direct quantitative relationship between the functional properties of human and macaque V5. Nat Neurosci 3(7):716–723. doi:10.1038/76673

Roberts TP, Mikulis D (2007) Neuro MR: principles. J Magn Reson Imaging 26(4):823–837. doi:10.1002/jmri.21029

Rueckert D, Sonoda LI, Hayes C, Hill DL, Leach MO, Hawkes DJ (1999) Nonrigid registration using free-form deformations: application to breast MR images. IEEE Trans Med Imaging 18(8):712–721

Schmierer K, Scaravilli F, Altmann DR, Barker GJ, Miller DH (2004) Magnetization transfer ratio and myelin in postmortem multiple sclerosis brain. Ann Neurol 56(3):407–415

Schmierer K, Wheeler-Kingshott CA, Tozer DJ, Boulby PA, Parkes HG, Yousry TA, Scaravilli F, Barker GJ, Tofts PS, Miller DH (2008) Quantitative magnetic resonance of postmortem multiple sclerosis brain before and after fixation. Magn Reson Med 59(2):268–277. doi:10.1002/mrm.21487

Sheehan DV, Lecrubier Y, Sheehan KH, Amorim P, Janavs J, Weiller E, Hergueta T, Baker R, Dunbar GC (1998) The Mini-International Neuropsychiatric Interview (M.I.N.I.): the development and validation of a structured diagnostic psychiatric interview for DSM-IV and ICD-10. J Clin Psychiatry 59(20):22–33 quiz 34-57

Sheehan DV, Sheehan KH, Shytle RD, Janavs J, Bannon Y, Rogers JE, Milo KM, Stock SL, Wilkinson B (2010) Reliability and validity of the Mini International Neuropsychiatric Interview for Children and Adolescents (MINI-KID). J Clin Psychiatry 71(3):313–326. doi:10.4088/JCP.09m05305whi

Shrout PE, Fleiss JL (1979) Intraclass correlations: uses in assessing rater reliability. Psychol Bull 86(2):420–428

Smith SM, Jenkinson M, Johansen-Berg H, Rueckert D, Nichols TE, Mackay CE et al (2006) Tract-based spatial statistics: voxelwise analysis of multi-subject diffusion data. NeuroImage 31:1487–1505

Thompson PM, Toga AW (1997) Detection, visualization and animation of abnormal anatomic structure with a deformable probabilistic brain atlas based on random vector field transformations. Med Image Anal 1(4):271–294

Thompson PM, Giedd JN, Woods RP, MacDonald D, Evans AC, Toga AW (2000) Growth patterns in the developing brain detected by using continuum mechanical tensor maps. Nature 404(6774):190–193

Toro R, Fox PT, Paus T (2008a) Functional coactivation map of the human brain. Cereb Cortex 18(11):2553–2559

Toro R, Perron M, Pike B, Richer L, Veillette S, Pausova Z, Paus T (2008b) Brain size and folding of the human cerebral cortex. Cereb Cortex 18(10):2352–2357

Toro R, Chupin M, Garnero L, Leonard G, Perron M, Pike B, Pitiot A, Richer L, Veillette S, Pausova Z, Paus T (2009) Brain volumes and Val66Met polymorphism of the BDNF gene: local or global effects? Brain Struct Funct 213(6):501–509

Upadhyay J, Hallock K, Ducros M, Kim DS, Ronen I (2008) Diffusion tensor spectroscopy and imaging of the arcuate fasciculus. Neuroimage 39(1):1–9. doi:10.1016/j.neuroimage.2007.08.046 S1053-8119(07)00775-6 [pii]

Zijdenbos AP, Forghani R, Evans AC (2002) Automatic "pipeline" analysis of 3-D MRI data for clinical trials: application to multiple sclerosis. IEEE Trans Med Imaging 21(10):1280–1291

Zilles K, Armstrong E, Schleicher A, Kretschmann HJ (1988) The human pattern of gyrification in the cerebral cortex. Anat Embryol (Berl) 179(2):173–179

Chapter 8
Cohorts

The previous chapters provided an overview of concepts and methodologies employed in acquiring information about an individual's envirome, genome and epigenome, and (brain) phenome. Population neuroscience attempts to bring this foundational knowledge together and apply it in the context of large, population-based studies. We need the population approach in order to accommodate the truly innumerable permutations of environmental and genetic factors that influence the brain phenotypes as they unfold from conception to death. Ideally, if we are to identify the most *common* pathways underlying cognitive abilities and mental health, then we need a large and representative sample of the *general population*. In this chapter, we begin by touching upon the most common study designs used in epidemiology. This is followed by specific examples as implemented in several current population-based studies of brain development. We close with a new proposal for future studies of brain health and cognitive decline that are carried out across multiple generations and integrate phenotypes acquired in multiple (brain and body) domains.

8.1 Case-Control and Cohort Studies: Main Distinctions

We typically use **case-control studies** when investigating differences in exposures between individuals diagnosed with a particular disorder (the cases) and those without it (the controls). This outcome-based design represents an efficient way for addressing relevant questions about a single disorder.

Alzheimer's disease is a good case in point. We may ask whether female patients with Alzheimer's disease (AD) differ from controls in their cumulative exposure to sex hormones (e.g. through their use of hormone replacement therapy, HRT). Thus, we recruit women diagnosed with AD (the cases) and matched to them, say by age, healthy women (the controls); then we ask whether there is a difference in the HRT use between the two groups. For several reasons, however, the use of such an a priori selection criterion limits the usefulness of this design for answering any other questions. By definition, case-control studies cannot be representative of the

T. Paus, *Population Neuroscience*, DOI: 10.1007/978-3-642-36450-1_8,

general population. Furthermore, unless a great care is taken in matching cases and controls on a number of possible confounders (e.g. neighbourhood, education, income), cases and controls may, in fact, come from different strata of the general population (Text Box 8.1.). Such methodological issues must be addressed carefully in case-control MR studies, such as the AD neuroimaging initiative (ADNI).

Text Box 8.1. Ascertainment bias

Ascertainment (to ascertain ~ to discover) refers to the way we identify individuals for the purpose of recruiting them into a given study. Ascertainment bias refers to a sampling (or selection) bias. Imagine a scenario where the recruitment of cases is carried out through a tertiary-care hospital (or a specialized clinic) but controls are recruited by local advertisements. Involvement of a state-of-the-art medical technology, such as MRI, may introduce additional biases by influencing the response rate: it is likely that the response rate would be higher in cases than controls, thus opening room for a selection bias (e.g. higher education and income in healthy individuals who decided to participate). In some cases, such a participation bias may, in fact, correlate with the exposure of interest (e.g. the more common use of hormonal replacement therapy in women with higher income).

It is important to note that the main goal of population neuroscience is to understand the *pathways* leading to health (or disease). Therefore, it is essential to begin such studies at a pre-clinical stage, free of confounders generated by the disease and its treatment, hence the cohort studies.

Cohort studies come in two flavours with regard to incorporating (or not) a particular exposure. If they do, the matter of whether or not an individual has been exposed to a particular environment (e.g. the use of HRT) drives the ascertainment. Obviously, the main advantage of this strategy lies in the efficiency of recruiting "exposed" participants, when the exposure affects less than half of the general population. Note, however, that—in this case—a particular (disease) outcome (e.g. AD) does not factor in during the recruitment; in fact, the full-blown disease may be years or decades away. Furthermore, recruitment of the "exposed" and "non-exposed" participants typically involves the same strategy (e.g. advertisement) and, as such, is less prone to biases associated with case-control studies, where the recruitment often differs between the cases (patients) and healthy controls. For these reasons, cohort studies designed in such a targeted way do not limit the breadth of questions about outcomes and the mechanistic pathways that lead to them; the only limitation is the representativeness of the sample vis-à-vis the general population. Nonetheless, the same caveat holds for the importance of matching "exposed" and "non-exposed" participants on possible confounders. We will discuss this below in the context of the Saguenay Youth Study Cohort, in which we ascertained (selected) adolescent participants based on their prenatal exposure to maternal cigarette smoking.

Finally, cohort studies that are unbiased by either exposures or outcomes during the recruitment stage could represent truly representative samples of the general population. As such, they afford the greatest flexibility in investigating the relationships between exposures and outcomes occurring—in a given cohort—with frequencies unbiased by ascertainment. Such cohorts, of course, can be established at any point in life, from conception to ageing. From a developmental perspective, however, young cohorts in general and birth cohorts in particular represent ideal opportunities for investigating environmental and genetic influences on the brain and behaviour as they unfold during the formative years. We will describe a number of such cohorts in the next section.

8.2 MR-Imaging in Cohort Studies

Over the past 20 years, MR-based imaging of the brain has become the tool of choice in a number of cohorts (Table 8.1).

Table 8.1 Population-based studies with a significant MR component

Study	Age (years)	MRI Brain
+ALSPAC	18–21	500
+Generation R: Wave 1	5–7	1,000
+Generation R: Wave 2	10–12	5,000*
IMAGEN	13.5–15.5	2,000
+NFBC1986	25–27	900
NIH-PD	0–18	500
NIMH-CHPB	3–25	400
Saguenay Youth Study	12–18	1,024
AGES (REYKJAVIK)	>70	5,000
ARIC	55–74	1,700
Framingham Heart	38–88	2,500
Health ABC	70–79	500
+Lothian birth cohort	>70	700
MESA	45–84	1,000
PURE-MIND	60	800
Rotterdam Study	60–90	5,000
Saguenay Parent Study	40–60	900

AGES Age, Gene/Environment Susceptibility; *ALSPAC* Avon Longitudinal Study of Parents and Children; *ARIC* Atherosclerosis Risk in Communities; *HEALTH ABC* Health Aging and Body Composition; *NFBC1986* Northern Finland Birth Cohort 1986; *MESA* Multi-Ethnic Study of Atherosclerosis; *NIH-PD* National Institutes of Health—Pediatric Database; *NIMH-CHP* National Institute of Mental Health—Child Psychiatry Branch; *PURE-MIND* Prospective Urban Rural Epidemiological Study. The last columns indicate the number of scanned participants. + Birth cohort, *Funded, scanning has not started yet

Given our interest in the developmental origin of health and disease, I will illustrate these by describing seven cohorts, spanning the first three decades of life (Table 8.1, top).

As described below, in some cases, the acquisition of MR images was the sole purpose for establishing a cohort (Sect. 8.2.1); naturally, with the variety of motivations behind these efforts comes a variety of designs (Sect. 8.2.2). In other instances, MR was used in a subset of participants of an already existing birth cohort (Sect. 8.2.3).

8.2.1 MR-Based Cohorts

Let me begin by describing four developmental cohorts designed right from the outset as MR-based. I will do so in the order in which they started their data collection; this chronology is reflected, to some extent, in their design, as summarized in Table 8.2.

8.2.1.1 National Institutes of Mental Health—Child Psychiatry Branch Cohort

This cohort was set up in 1989 as a normative study of brain structure during childhood and adolescence. It has been carried out at one acquisition site: Bethesda, Maryland, U.S.A. One of the primary goals of the study has been the comparison of normative data with MR images acquired in parallel studies of childhood psychiatric disorders, including conditions both common (e.g. attention deficit hyperactivity disorder, ADHD) and rare (e.g. childhood-onset schizophrenia). It is a longitudinal study, with the participants' ages at the time of initial recruitment ranging from 3 to 25 years, and the visits repeated in two- to four-year intervals (e.g. Giedd et al. 1996; Lenroot et al. 2007). Given that this was the very first large-scale MR study of brain structure in typically developing children and adolescents, it was designed primarily as an MR study, with little data collected on genetic and environmental exposures other than the socioeconomic status (SES) of the families; DNA collection was added at a later point. Behavioural and cognitive assessments are not extensive (see Table 8.2, first column). The neuroimaging protocol includes T1-weighted images: DTI and functional MRI were added at later stages in a subset of participants. The sample comes primarily from the local community. In terms of the recruitment strategy, this study has relied to a great extent on the interest of NIH employees working on the Bethesda campus. As a consequence, this is a multi-ethnic sample, with an average estimated IQ of 113.

Table 8.2 An overview of four population-based studies of brain development

	NIMH-CHPB	NIH-PD	SYS	IMAGEN
Sample (n)	400	500	1,024	2,000
Age range	3–25 years	7 days–18 years	12–18 years	14 years
Design	Longitudinal, multi-ethnic population	Longitudinal, multi-ethnic population	Cross-sectional, founder population	Cross-sectional, multi-ethnic population
Recruitment	Local community	Population sampling (census-based targets)	High schools	High schools
Genetics	Candidate genes	None	DNA (adolescents, biological parents)	DNA (adolescents)
Environment	Socioeconomic status	Socioeconomic status	Socioeconomic status	Socioeconomic status
		Pregnancy (smoking, alcohol, drugs)	Pregnancy (smoking, alcohol, drugs)	Pregnancy (smoking, alcohol, drugs)
			Infancy (breast-feeding)	Stressful life events
			Childhood (maternal care, stressful life events, food availability/variety)	
			Adolescents (diet, sleep)	
Imaging	T1W images, 1.5-mm-thick axial slices, 1.5T GE scanner	T1W, T2 W, PDW images, T1 and T2 relaxometry, 1.5T GE and Siemens scanners	Brain: T1W, T2W, PDW images, MTR; Abdomen: fat, kidney volume; 1.0T Philips scanner	Structural: T1W images, DTI; functional: face task, MID task, stop-signal task, global-cognition Task; 3-T scanners (GE, Philips, Siemens, Bruker)
Behaviour/ Cognition	Child and Parent diagnostic (psychiatric) Interview for children	DISC	DISC Predictive Scale (psychiatric symptoms)	DAWBA and SDQ (psychiatric symptoms)
	Child behaviour checklist	Child behaviour checklist	Positive youth development, personality (NEO-PI), Anti-social behaviour	Personality (NEO FFI, TCI-R)

(continued)

Table 8.2 (continued)

	NIMH-CHPB	NIH-PD	SYS	IMAGEN
	Intelligence (WISC-III/ WISC-IV subtests)	Personality (TCI)	Drug experimentation, Sleep, Sexuality	Substance use (SUPRS, ESPAD, DAST, AUDIT, MAST, FTND, TLFB)
	Spatial working memory, Go/No-Go task	Intelligence (WASI, WISC-III subtests)	Intelligence (WISC-III)	Intelligence (WISC subscales)
	Academic skills (reading, spelling)	Memory (CVLT)	Memory (CMS)	Executive functions (CANTAB)
	Grooved Pegboard and Handedness	Executive functions (CANTAB, NEPSY)	Executive functions (e.g. Stroop, Fluency, Working Memory, Attention)	Face perception
		Academic skills (calculation, passage comprehension, letter word)	Face perception, body-image perception	Handedness and Fine motor skills
		Handedness and Fine motor skills	Reward/Impulsivity	
			Phonological processing	
			Academic skills (math, math fluency, reading, spelling) and number sense	
			Handedness and Fine motor skills	

National Institutes of Health—Child Psychiatry Branch (NIH-CHPB); National Institutes of Health—Pediatric Database (NIH-PD); Saguenay Youth Study (SYS).
T1W, T1-weighted; T2W, T2-weighted; Proton density-weighted (PDW); Magnetization transfer ratio (MTR); monetary incentive delay (MID); general electrics (GE); Diagnostic interview schedule for children (DISC); development and well-being assessment interview (DAWBA); strengths and difficulties questionnaire (SDQ); substance use risk profile scale (SUPRS); European school survey project on alcohol and drugs (ESPAD); drug abuse screening test (DAST); alcohol use disorders identification tests (AUDIT); Michigan alcohol screening test (MAST); Fagerstrom test for nicotine dependence (FTND); time follow-back interview (TLFB); neuroticism-extroversion-openness personality inventory (NEO-PI); neuroticism-extroversion-openness five factor inventory (NEO FFI); temperament and character inventory (TCI); Wechsler abbreviated scale of intelligence (WASI); Wechsler intelligence scale for children (WISC); California verbal learning test (CVLT); children's memory scale (CMS); Cambridge neuropsychological test automated battery (CANTAB); A Developmental NEuroPSYchological assessment (NEPSY).
From Paus (2010).

8.2.1.2 National Institutes of Health—Pediatric Database

This project was initiated in the mid-2000 as a multi-centre study where MR acquisition took place at six sites in the U S A. It is also a normative study of brain development that complements the NIH-CHB cohort in two important ways: (1) by including a large group of infants and young children (age 7 days–4 years) and (2) by adding other (structural) MR sequences (Evans 2006). It is a longitudinal study, with the older children (5–18 years) scanned three times in two-year intervals and the younger children (7 days–4 years) scanned up to five times, with intervals as short as 3 months. No genetic data were collected.

Assessment of environmental exposures was limited to variables related to SES and prenatal exposures (e.g. cigarettes, alcohol and drugs). Behavioural and cognitive assessments were more extensive than those in the NIH-CHB cohort (see Table 8.2, second column and Waber et al. 2007). The NIH-PD project included several MR sequences: T1-, T2- and PD-weighted images, as well as T1 and T2 (single slice) relaxometry, were acquired in all participants. Diffusion tensor images and MR spectroscopy were acquired at a subset of the acquisition sites.

The sample was ascertained through population-based sampling: each of the six acquisition sites recruited participants using site-specific demographic targets calculated according to the U.S. Census 2000 data. The resulting sample is multi-ethnic and includes a wide range of SES characteristics. The average IQ is 110.

8.2.1.3 Saguenay Youth Study

The Saguenay Youth Study was initiated in the mid-2000 as an investigation of genetic and environmental factors shaping the adolescent brain and body. MR acquisition took place at a single site in Canada. Adolescents (12–18 years) and their biological parents were recruited from a population with a known genetic founder effect (Text Box 8.2.), in the population of the Saguenay Lac-Saint-Jean (SLSJ) region of Quebec, Canada.

Text Box 8.2. Genetic founder effect

Founder populations provide important advantages for genetic and epidemiological research. Compared with the outbred populations constituting most of today's migratory world, they are more homogenous in genetic background and, due to their relative geographical isolation, also in environmental exposures such as cultural habits, diet, climate (Peltonen et al. 2000). Because of this genetic homogeneity, fewer genes and gene variants are thought to contribute to the phenotypic expression of complex genetic traits (De Braekeleer 1991). This founder effect, combined with more uniform environmental exposures, is expected to increase the likelihood of identifying genes underlying complex genetic traits.

At its inception, this was a cross-sectional study. Multiple quantitative phenotypes relevant to mental, cardiovascular and metabolic health were acquired, using an extensive 15 h protocol spread over several days (Pausova et al. 2007). A follow-up of the participants will begin in 2014. By design, half of the participants were exposed to maternal cigarette smoking (the cases), while the other, non-exposed half has been matched to them by maternal education (the controls). A family-based design was used, with adolescent siblings fully phenotyped. Their biological parents provided only a blood sample for genetic analyses and answered questions about their current mental health. (The full MR-based assessment of the parents is in progress).

Recruitment took place in high schools across the SLSJ region. Samples of DNA were collected in all adolescents and their biological parents. Assessment of environmental exposures covered the prenatal period (e.g. smoking, alcohol), infancy (e.g. breastfeeding), childhood (e.g. food availability, maternal care, stressful life events) and adolescence (e.g. diet, sleep). Behavioural and cognitive assessments were extensive and included both self-reported psychiatric symptoms, components of positive youth development and personality, as well as a thorough 6 hour assessment of cognitive abilities (see Table 8.2, third column and Pausova et al. 2007; Kafouri et al. 2009). In addition, all adolescents were assessed with a detailed cardiovascular and metabolic protocol. MR sequences included T1-weighted, T2-weighted and PD-weighted images, magnetization transfer (MT) images (as an index of myelination) and abdominal images (extra- and intraabdominal fat, kidney volume). The sample is of a single ethnicity (white Caucasians) and, given the 50 % inclusion rate of adolescents born to mothers who smoked during pregnancy and the matching procedure, it is of lower SES than the general population of the region. The average IQ is 105.

8.2.1.4 IMAGEN

This project started in 2007 as a multi-centre, cross-sectional study of the genetic and neurobiological bases of individual variability in impulsivity, reinforcer sensitivity and emotional reactivity. MR acquisition took place at eight acquisition sites located in the United Kingdom, Ireland, France and Germany (Schumann et al. 2010). Adolescents (14 years old) were recruited primarily through local high schools. Samples of DNA were collected in all participating adolescents. Assessment of environmental exposures was limited to the main SES characteristics, stressful life events and prenatal exposures (e.g. smoking, alcohol). Behavioural and cognitive assessments included a basic assessment of cognition and a detailed assessment of the main outcomes of interest, namely impulsivity, reward processing and substance use (see Table 8.2, fourth column). MR sequences included structural (T1-weighted, DTI) and functional imaging, with the latter consisting of four paradigms: a Face Task, a Monetary Incentive Delay (MID) Task, a Stop-Signal Task and a Global-Cognition Task. At 16 years of age, about 80 % of the original sample completed a web-based follow-up of mental health and substance use. At 18 years of age, all adolescents will be invited for a follow-up visit that will include the same structural and functional MR imaging acquired at 14 years of age. The sample is multi-ethnic, with a wide range of the parental education level. The average estimated IQ is 108.

8.2.2 Comparison of the Four Cohorts: Recruitment and Assessments

Given the variety of primary goals and the time of inception, these four studies of brain development range significantly in their approaches to the selection and recruitment of typically developing children and/or adolescents from the local population, the collection of genetic and environmental variables and the inclusion of specific MR sequences.

In the case of **selection and recruitment**, the NIH-CHB and NIH-PD studies represent two extremes, with the former being a convenience sample and the latter a stratified random sample. It should be pointed out, however, that using a random-sampling strategy does not necessarily yield a random sample in the final dataset. This is chiefly due to two reasons. First, given the volunteer nature of neuroimaging studies, there is a self-selection bias in the initial stage of recruitment. Second, depending on the stringency of the exclusion criteria, a varied number of potential participants are excluded from the final sample, thus creating a "super-healthy" cohort. In this respect, the NIH-CHB and NIH-PD studies are more exclusionary than the SYS and IMAGEN studies. Note that this strategy may reduce the heterogeneity of the sample vis-à-vis exposures associated with negative outcomes in early age. Overall, the true randomness and representativeness of population-based samples are difficult to achieve in studies requiring significant commitment of the participants; with adequate resources, this limitation can be mitigated to some extent by compensating participants for their time and inconvenience.

The **availability of DNA** and, therefore, the potential for genetic analyses vary across the four studies—from no DNA having been collected in the NIH-PD study, to DNA being collected from adolescent participants and their biological parents in the SYS cohort. Furthermore, it is possible that the multi-ethnic population of the NIH-CHB and the IMAGEN studies would require a larger sample size in order to reveal particular genotype–phenotype relationships, as compared with the single ethnicity of the SYS cohort recruited from a population with a known genetic founder effect (Kristiansson et al. 2008; see also Text Box 8.2).

As seen in Table 8.2, assessment of the **environment** varied widely across the studies. All four studies collected basic SES characteristics, which may serve as a proxy of the "social context of early life" (Taylor et al. 2004) and a possible marker of the "stressfulness" of the family environment (Adler et al. 1999). All but one of the studies documented a possible exposure of the foetus to maternal cigarette smoking and alcohol/drug use. Two studies collected information about stressful life events and one of the studies extended environmental assessment to other variables, such as breast-feeding, food availability and type of care during childhood. Overall, assessment of environmental exposures relies mostly on parental recall; it is rarely verified using independent sources (e.g. medical records) and often attempts to capture events that occurred in the relatively distant past, such as during pregnancy.

When it comes to **behavioural and cognitive phenotypes**, all four studies assessed current psychiatric symptoms—using self-reports or structured interviews—and general intelligence, the latter estimated from a limited number of subscales in the NIH-CHB and the IMAGEN studies. Three studies (NIH-PD, IMAGEN, SYS) used the NEO-PI and/or TCI to assess personality. Two of the studies (NIH-PD and IMAGEN) used various subtests of CANTAB to assess executive functions. Being focused on adolescence, the SYS and the IMAGEN studies include an extensive assessment of substance use.

Finally, we have **the brain phenotypes**. All fours studies include T1-weighted images of the brain, which provide a wealth of quantitative phenotypes (see Chap. 7). The IMAGEN Study and, in part, the NIH-PD dataset also acquired DTI images, which allow for the characterization of white-matter microstructure. The SYS includes acquisition of MT images used to calculate an MR ratio (MTR), an indirect index of myelination. Finally, the IMAGEN Study is the only of the four studies to collect functional MR datasets quantifying, respectively, the brain response to social stimuli (the Face task), processing of rewards (the MID Task) and response inhibition (the Stop-Signal Task).

8.2.3 Birth Cohorts with an MR Arm

Let me now turn attention to three birth cohorts and their approach to including MRI in their designs.

The first two cohorts started between 20 and 26 years ago, without having included an MRI protocol from the outset. This has been added only at a later point in a subset of young (18- to 26-year-old) cohort members for specific reasons, as described below. In these two cases, the additional collection of MR-based phenotypes will be integrated with the existing longitudinal data acquired between pregnancy and the present. This is the key advantage of adding MR imaging in an ongoing birth cohort at any point. The third, more recent, cohort has taken the opportunity to add a substantial MR-based assessment into its overall design in its early stages (5–12 years of age).

8.2.3.1 Avon Longitudinal Study of Parents and Children

This is a birth cohort ascertained in southwest England between 1 April 1991 and 31 December 1992. The initial cohort included 13,971 children alive at 12 months of age. Between 2010 and 2012, we acquired brain MRIs in 507 male participants in order to investigate the relationship between testosterone trajectories during puberty and the structural properties of white mater in late adolescence/early adulthood.

All 10,000 ALSPAC participants have been tested at multiple points during their infancy, childhood and adolescence. Detailed information has been collected, from pregnancy onwards, using self-administered questionnaires and research clinics lasting up to 3 h. The two key instruments used in ALSPAC to evaluate psychopathology during childhood include the Development and Well-Being Assessment (Goodman et al. 2000a) and Strengths and Difficulties Questionnaire (Goodman et al. 2000b). A number of additional instruments have been used to assess anti-social behaviour, psychosis-like symptoms, depressive symptoms and drug-related behaviour. Cognitive abilities have been tested repeatedly (at different ages) using a variety of instruments, including WISC-III, WASI, digit span, attention and Stop-Signal Task. DNA is also available for all participants. Blood samples—for the assessment of testosterone levels—have been collected at 9, 11, 13, 15 and 17 years of age (three to five samples available in each participant). Together with the longitudinal data about the participants' psychopathology, this detailed characterization of each participant during his childhood and adolescence will allow us to identify the key predictors of brain health, as assessed with MRI in young adulthood.

8.2.3.2 Northern Finland Birth Cohort 1986

This birth cohort was established as a prospective longitudinal cohort of mothers living in the two northern-most provinces of Finland, Oulu and Lapland, with expected dates of delivery falling between 1 July 1985 and 30 June 30 1986. The cohort includes a total of 9,432 live births. Interviews and postal questionnaires were completed/returned from the 24th gestational week onwards. In 2011, we began acquisition of brain MR images in 900 25 to 27-year-old NFBC86 participants (50 % exposed to maternal smoking during pregnancy).

In all 9,000 NFBC86 participants, the course of the pregnancy and delivery, including complications, was confirmed using patient records, as was the neonatal outcome. The children were followed up at the ages of 6–12 months (via special nursing cards), 7–8 years and 14–16 years. These follow-ups included assessments of psychopathology in childhood by their parents and teachers, as well as self-assessments of psychopathology and substance use in adolescence. DNA is available in all participants.

The design of the MR arm of this cohort parallels that of the Saguenay Youth Study. The slightly higher age of the NFBC86 participants at the time of imaging (26 years vs. the average of 14.5 years in the SYS) will allow us to investigate brain-behaviour relationships relevant for substance use, and their moderation by genes and exposure, at the age when substance use peaks in frequency (Chen and Jacobsen 2012). Given the longitudinal nature of the cohort and an excellent linkage of health-care databases available in Finland, we will be also in a strong position to identify—in a prospective manner—brain predictors of substance use and mental health later in life.

8.2.3.3 Generation R

This birth cohort is a prospective cohort study established between April 2002 and January 2006 in a multi-ethnic population of Rotterdam, the Netherlands. In total, 9,778 mothers were enrolled in the study during this period; 7,893 children are available for follow-up studies. Extensive data have been collected during pregnancy and preschool period (birth to 4 years of age), including physical and ultrasound examinations, questionnaire-based assessments of behaviour and cognition, and biological samples (Jaddoe et al. 2010). From the age of 5 years onwards, children visit a research centre for detailed measurements relevant for a number of domains, including respiratory and cardiovascular health, obesity, eye and tooth development, as well as mental health and cognition.

In September 2009, the Generation R team initiated the first wave of MR-based assessments of brain development, with the target of 1,000 children scanned between 5 and 7 years of age. The imaging is carried out on a single 3-T scanner (GE Discovery 750) with a 40-min protocol that includes the following sequences: T1-weighted and PD-weighted scans, DTI and resting fMRI.

In 2013, the team will start the second wave of MR imaging, with the target of over 5,000 children to be scanned between 10 and 12 years of age. The scanning will be carried out on a new 3-T scanner (GE Discovery 750 wide-bore) dedicated solely to this purpose. The planned 30-min protocol will include the following sequences: T1-weighted images, DTI, field map and resting fMRI. Clearly, the inclusion of the MR arm in this birth cohort represents a unique opportunity for large-scale integration of state-of-the-art longitudinal assessments obtained in this cohort (e.g. foetal ultrasound-based brain imaging, parent–child interaction, physical and social environment), with a detailed MR-based evaluation of structural and functional properties of the developing brain. Again, this information will be invaluable in identifying brain predictors of mental and cognitive health, as the cohort members move through adolescence to adulthood.

8.3 A Trans-Generational Brain and Body Cohort

Let us assume that we are interested in identifying predictors of successful (brain) ageing. What do we need to do to get to an advanced age unimpaired by cognitive decline and/or dementia? As we explain in Chap. 10, one possible answer is personalized preventive medicine. To reach this long-term goal, we must be well equipped to make long-term predictions about how genetic and environmental risk factors affect brain health. We argue that this is best achieved if the following two elements are kept in mind: (1) the cohort is trans-generational; and (2) the cohort integrates phenotypic information about both the brain and body.

8.3.1 The Trans-Generational Element

The trans-generational element represents a possible "short cut" for achieving (and validating) long-term predictions. Ideally, we would design a longitudinal study to span from cradle to grave. Many birth cohorts have this ambition in mind and some, such as the Lothian Birth Cohort, are reaching the end points. But if one wishes to use modern technology in such studies, the length of the human life works against us, because what is state-of-the-art today will be obsolete (and unavailable) tomorrow. To overcome this challenge, we can design instead a multi-generational study, where the successive generations represent an approximation of "what might happen" in the future ("like father, like son"). In other words, the trans-generational commonalities—based on the shared genes and (to some extent) the shared family environment—become a "signature" of a given family on which non-shared elements (e.g. individual lifestyle, treatments) operate. By comparing individuals of the same age, but coming from families with the different "signatures," we may be in a position to identify long-term predictors of brain health.

Thus, we hypothesize that the accuracy of discriminating between a descendant (e.g. daughter or granddaughter) who will develop a disease and one who will not, tested against the profile (and/or disease status) of her ancestor (e.g. mother or grandmother), will be comparable to the discriminative accuracy observed in recent epidemiological studies on a short time-scale. As shown by our collaborators in the case of mortality risk in the Rotterdam Study, the predictive value of a number of lifestyle and physiological characteristics (162 variables in total) is high vis-à-vis the short-term (<1 year) prediction of mortality ($\sim$0.80) but it decreases to $\sim$0.70 when death occurs 15 years post-assessment (Henning Tiemeier, Stefan Walter; personal communication, June 21, 2011). This is illustrated in the top part of Fig. 8.1 (dashed lines). If our hypothesis is confirmed, then we posit that Prognosis by Ancestor/Pedigree may be equal to or better than a long-term (30+ years) prognosis by one's current health status (bottom part of Fig. 8.1, solid lines). If so, Prognosis by Ancestor/Pedigree will provide a glimpse into the future for descendants and the basis for taking specific interventions and preventive measures in the case of a poor prognosis, so as to avoid having it come to pass.

8.3.2 The Integrated Brain and Body Element

The integrated brain and body element is of the essence due to the complexity of mechanistic pathways leading to dementia and their relationship to cardiovascular and metabolic health. We know, for example, that metabolic syndrome, a cluster

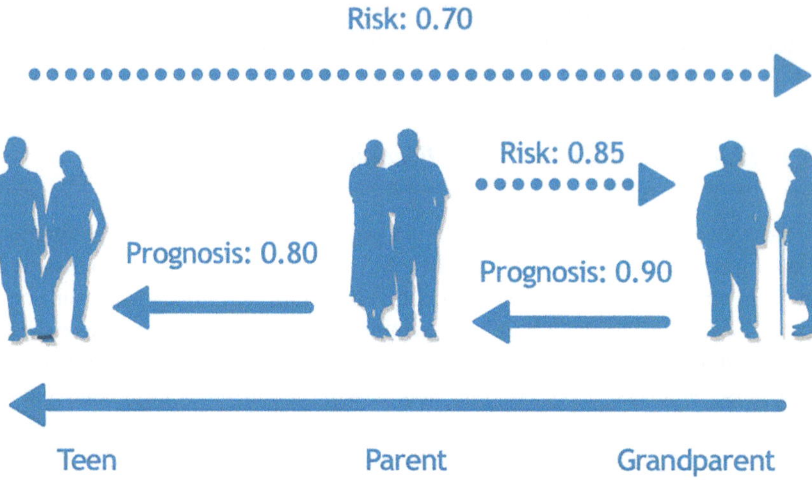

Fig. 8.1 Risk/resilience profiling: virtual validation (*dashed lines*) and prognosis by ancestor/pedigree (*solid lines*). The numbers indicate hypothesized discriminative accuracy (0.5 = chance, 1 = perfect discrimination)

of risk factors for cardiovascular diseases and type 2 diabetes, is associated with a 2.5-fold increase in the risk of AD (Vanhanen et al. 2006; see also Kivipelto et al. 2006 and Frisardi et al. 2010). Obesity is clearly one of the key components of the metabolic syndrome, with negative consequences on both cardiovascular (high blood pressure, atherosclerosis) and metabolic (insulin resistance) health. But not all body fat is equally dangerous: fat surrounding internal organs in the abdominal cavity, the so-called visceral fat (Text Box 8.3), appears to be particularly harmful vis-à-vis brain health (Anan et al. 2010; Benedict et al. 2012; Debette et al. 2010; Isaac et al. 2011).

Text Box 8.3 Visceral fat and its toxicity

Visceral fat may be more harmful with regard to physical and cognitive health for a number of reasons. First, there is a greater flux of lipids from visceral fat than from any other fat depot; visceral (vs. subcutaneous) fat exhibits greater lipid turnover, including higher fat uptake and lypolysis (Votruba and Jensen 2007). Second, visceral fat drains directly into the portal circulation and the liver, where excess free fatty acids can contribute to the enhanced *de novo* synthesis of triacylglycerols, which are released into the circulation bound in very-low-density lipoproteins (Votruba and Jense 2007); these may, in turn, increase sympathetic activity and blood pressure (Haastrup et al. 1998; Florian and Pawelczyk 2010a, b; Grekin et al. 1995; Stojiljkovic et al. 2001). Third, free fatty acids and various lipid species (e.g. phospholipids, sphingolipids) may also affect directly a number of brain processes through their participation in various signalling pathways (e.g. apoptosis, oxidative stress, inflammation; Farooqui et al. 2007, 2012; Narayan and Thomas 2011), myelination (Vyas and Schnaar 2001), and neurotransmission (e.g. through "chaperone-like" effects of lipid rafts on neurotransmitter receptors; Tsui-Pierchala et al. 2002; Fantini and Barrantes 2009).

This is one of the reasons why we have incorporated MR-based quantification of visceral fat into the Saguenay Youth Study (see Sect. 8.2.1.3), and especially in the Saguenay Parent Study. Together, the Saguenay Youth Study and the Saguenay Parent Study represent the first multi-generational cohort assessed with the same phenotyping tools, including the brain and abdominal MRI. In both cohorts, we also devote over 1 h to a detailed assessment of blood pressure and its regulation (Fig. 8.2). We do so using a non-invasive hemodynamic monitor Finometer™ (FMS Finapres, Amsterdam, The Netherlands), a device that measures finger blood flow continuously and, from these data, derives beat-by-beat brachial blood pressure using the reconstruction and level correction of the finger blood-flow waveform. The Finometer™ is a reliable device for tracking blood pressure in adults and children older than 6 years (Tanaka et al. 1994). The following measures are calculated beat-by-beat: systolic and diastolic blood pressure, inter-beat interval (heart rate), stroke volume, cardiac output and total peripheral resistance. In addition, we can calculate the power in a particular (low) frequency band contained in the time series of diastolic blood pressure and, from these, estimate sympathetic modulation of vasomotor tone (Pagani et al. 1986).

In the Saguenay Youth Study, we have shown that accumulation of visceral fat is associated with higher sympathetic modulation of the vasomotor tone and blood pressure in adolescent males but not in females (Syme et al. 2008) and that, also in males but not in females, these associations may be regulated by the CAG-repeat number in the coding region of the androgen receptor gene (Pausova et al. 2010; Fig. 8.3).

Overall, it is highly likely that we can improve long-term predictions of brain health by incorporating quantitative phenotypes of other key organs. In particular, owing to the intimate relationship between the cardiovascular and metabolic systems and their impact on brain structure and function, we should consider assessing the function and structure of the heart and the main vessels, as well as the amount and distribution of adipose tissue, when designing MR studies of brain health. For this reason, we have developed an integrated MR protocol (Table 8.3.) that includes—in a total of 90 min—cardiac imaging, imaging of body-fat distribution (Fig. 8.4) and ectopic fat in the liver, and brain imaging.

As we will discuss in Chap. 10, this type of phenotyping protocol is going to provide a rich system-level dataset, to be combined with molecular-level phenotypes as well as with the individual's envirome and genome. This will build his/her risk (and resilience) profile. Doing so in the trans-generational context provides another step in the development and validation of such profiles.

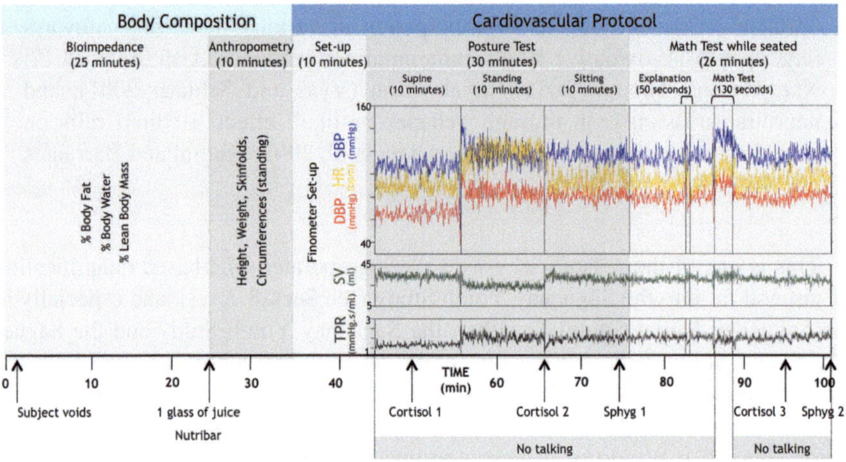

Fig. 8.2 Body-composition and cardiovascular protocol used in the Saguenay Youth Study. Blood pressure is measured at every heartbeat with Finometer. *SBP* systolic blood pressure; *DBP* diastolic blood pressure; *HR* hear rate; *SV* stroke volume; *TPR* total peripheral resistance; Sphyg 1 and 2, additional measurements of blood pressure using sphygmomanometer; Cortisol 1, 2 and 3, samples of saliva obtained for cortisol measurements. From Pausova et al. (2007)

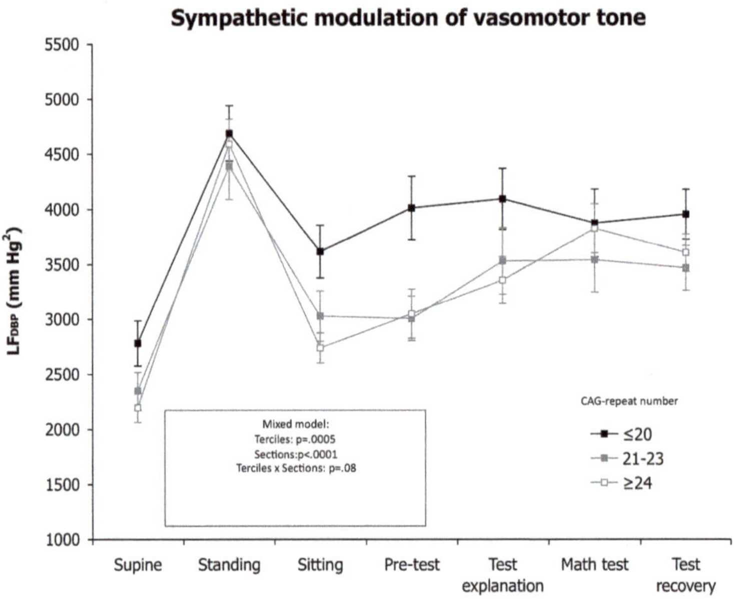

Fig. 8.3 Sympathetic activation (indexed by low frequency fluctuations in diastolic blood pressure; LF_{DBP}) in male adolescents with different number of cytosine-adenine-guanine (*CAG*) repeats in the androgen receptor gene (*AR*), displayed as a function of the cardiovascular protocol. Note that a low number of CAG repeats (<20) is associated with stronger trans-activational effects of androgens, that is, higher androgenicity. From Pausova et al. (2010)

Table 8.3 An integrated MRI protocol

MRI Sequence	Time (min)	Structure and physiology
PHASE 1: Cardiac, abdomen and liver spectroscopy (24 min)		
Heart: structure and function	8	End diastolic and systolic volumes, myocardial mass, ejection fraction, cardiac output, stroke volume
Blood flow	8	Blood flow in aorta, aortic wall stiffness
Myocardial T1 Mapping	1	Myocardial fibrosis
In–out phase abdomen	4	Visceral fat, kidney volume
Liver spectroscopy	3	Ectopic fat in liver
PHASE 2: Whole-body adiposity (16 min)		
Whole body	16	Adiposity, muscle (8 sections from head to toe)
PHASE 3: Brain (34 min)		
3D T1-weighted	5	Volumes, thickness, folding, shape, tissue density
Diffusion tensor imaging	10	Fractional anisotropy, mean diffusivity, track delineation
T2-weighted/ FLAIR	3	Hyperintensities (number, volume, location)
Magnetization transfer	8	Myelination index
Arterial spin labelling/BOLD	8	Perfusion/cerebro-vascular reactivity

Note that the protocol consists of three "phases"; participants take a short break (and stand up) between each phase. This protocol is currently running in the Toronto Trans-generational Brain and Body Study (www.brainbody.org); it takes a total of 90 min (with breaks and repositioning)

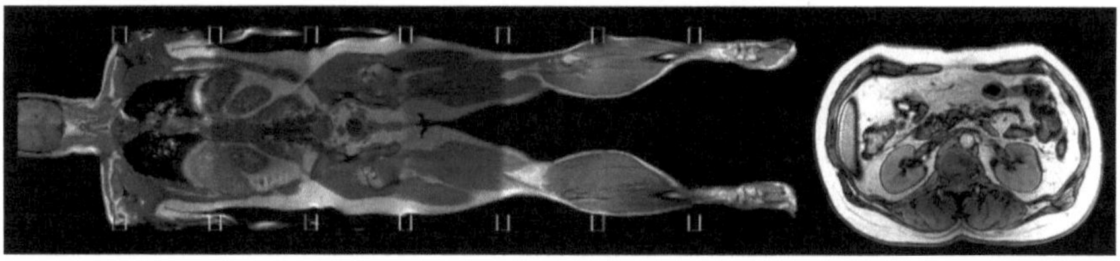

Fig. 8.4 Magnetic resonance images of whole-body adiposity and muscle mass (*left*) and subcutaneous and visceral fat (*right*)

In summary, MR-enriched cohort studies provide a unique opportunity for acquiring brain phenotypes at a preclinical level and, in turn, using this information to characterize trajectories, and molecular pathways, leading to a disease.

References

Adler NE, Marmot M, McEwen BS, Stewart J (eds) (1999) Socioeconomic status and health in industrial nations: Social, psychological, and biological pathways. Ann NY Acad Sci 896

Anan F, Masaki T, Shimomura T, Fujiki M, Umeno Y, Eshima N, Saikawa T, Yoshimatsu H (2010) Abdominal visceral fat accumulation is associated with hippocampus volume in non-dementia patients with type 2 diabetes mellitus. Neuroimage 49(1):57–62. doi:S1053-8119(09)00900-8 [pii] 10.1016/j.neuroimage.2009.08.021

Benedict C, Brooks SJ, Kullberg J, Burgos J, Kempton MJ, Nordenskjold R, Nylander R, Kilander L, Craft S, Larsson EM, Johansson L, Ahlstrom H, Lind L, Schioth HB (2012) Impaired insulin sensitivity as indexed by the HOMA score is associated with deficits in verbal fluency and temporal lobe gray matter volume in the elderly. Diabetes Care 35(3):488–494. doi:dc11-2075 [pii] 10.2337/dc11-2075

Chen P, Jacobson KC (2012) Developmental trajectories of substance use from early adolescence to young adulthood: gender and racial/ethnic differences. J Adolesc Health 50(2):154–163. doi:S1054-139X(11)00192-3 [pii] 10.1016/j.jadohealth.2011.05.013

De Braekeleer M (1991) Hereditary disorders in Saguenay-Lac-St-Jean (Quebec, Canada). Hum Hered 41(3):141–146

Debette S, Beiser A, Hoffmann U, Decarli C, O'Donnell CJ, Massaro JM, Au R, Himali JJ, Wolf PA, Fox CS, Seshadri S (2010) Visceral fat is associated with lower brain volume in healthy middle-aged adults. Ann Neurol 68(2):136–144. doi:10.1002/ana.22062

Evans AC (2006) The NIH MRI study of normal brain development. Neuroimage 30(1):184–202

Fantini J, Barrantes FJ (2009) Sphingolipid/cholesterol regulation of neurotransmitter receptor conformation and function. Biochim Biophys Acta 1788(11):2345–2361. doi:S0005-2736(09)00297-1 [pii] 10.1016/j.bbamem.2009.08.016

Farooqui AA, Farooqui T, Panza F, Frisardi V (2012) Metabolic syndrome as a risk factor for neurological disorders. Cell Mol Life Sci 69(5):741–762. doi:10.1007/s00018-011-0840-1

Farooqui AA, Horrocks LA, Farooqui T (2007) Interactions between neural membrane glycerophospholipid and sphingolipid mediators: a recipe for neural cell survival or suicide. J Neurosci Res 85(9):1834–1850. doi:10.1002/jnr.21268

Florian JP, Pawelczyk JA (2010a) Non-esterified fatty acids increase arterial pressure via central sympathetic activation in humans. Clin Sci (Lond) 118(1):61–69. doi:CS20090063 [pii] 10.1042/CS20090063

Florian JP, Pawelczyk JA (2010b) Sympathetic and haemodynamic responses to lipids in healthy human ageing. Exp Physiol 95(4):486–497. doi:expphysiol.2009.050997 [pii] 10.1113/expphysiol.2009.050997

Frisardi V, Solfrizzi V, Seripa D, Capurso C, Santamato A, Sancarlo D, Vendemiale G, Pilotto A, Panza F (2010) Metabolic-cognitive syndrome: a cross-talk between metabolic syndrome and Alzheimer's disease. Ageing Res Rev 9(4):399–417. doi:10.1016/j.arr.2010.04.007

Giedd JN, Snell JW, Lange N, Rajapakse JC, Casey BJ, Kozuch PL, Vaituzis AC, Vauss YC, Hamburger SD, Kaysen D, Rapoport JL (1996) Quantitative magnetic resonance imaging of human brain development: ages 4–18. Cereb Cortex 6(4):551–560

Goodman R, Ford T, Richards H, Gatward R, Meltzer H (2000a) The development and well-being assessment: description and initial validation of an integrated assessment of child and adolescent psychopathology. J Child Psychol Psychiatry 41(5):645–655

Goodman R, Ford T, Simmons H, Gatward R, Meltzer H (2000b) Using the strengths and difficulties questionnaire (SDQ) to screen for child psychiatric disorders in a community sample. Br J Psychiatry 177:534–539

Grekin RJ, Vollmer AP, Sider RS (1995) Pressor effects of portal venous oleate infusion. A proposed mechanism for obesity hypertension. Hypertension 26(1):193–198

Haastrup AT, Stepniakowski KT, Goodfriend TL, Egan BM (1998) Intralipid enhances alpha1-adrenergic receptor mediated pressor sensitivity. Hypertension 32(4):693–698

Isaac V, Sim S, Zheng H, Zagorodnov V, Tai ES, Chee M (2011) Adverse associations between visceral adiposity, brain structure, and cognitive performance in healthy elderly. Front Aging Neurosci 3:12. doi:10.3389/fnagi.2011.00012

Jaddoe VW, van Duijn CM, van der Heijden AJ, Mackenbach JP, Moll HA, Steegers EA, Tiemeier H, Uitterlinden AG, Verhulst FC, Hofman A (2010) The Generation R study: design and cohort update. Eur J Epidemiol 25(11):823–841. doi:10.1007/s10654-010-9516-7

Kafouri S, Leonard G, Perron M, Richer L, Seguin JR, Veillette S, Pausova Z, Paus T (2009) Maternal cigarette smoking during pregnancy and cognitive performance in adolescence. Int J Epidemiol 38(1):158–172

Kivipelto M, Ngandu T, Laatikainen T, Winblad B, Soininen H, Tuomilehto J (2006) Risk score for the prediction of dementia risk in 20 years among middle aged people: a longitudinal, population-based study. Lancet Neurol 5(9):735–741. doi:S1474-4422(06)70537-3 [pii] 10.1016/S1474-4422(06)70537-3

Kristiansson K, Naukkarinen J, Peltonen L (2008) Isolated populations and complex disease gene identification. Genome Biol 9(8):109. doi:gb-2008-9-8-109 [pii] 10.1186/gb-2008-9-8-109

Lenroot RK, Gogtay N, Greenstein DK, Wells EM, Wallace GL, Clasen LS, Blumenthal JD, Lerch J, Zijdenbos AP, Evans AC, Thompson PM, Giedd JN (2007) Sexual dimorphism of brain developmental trajectories during childhood and adolescence. Neuroimage 36(4):1065–1073

Narayan S, Thomas EA (2011) Sphingolipid abnormalities in psychiatric disorders: a missing link in pathology? Front Biosci 16:1797–1810. doi:3822 [pii]

Pagani M, Lombardi F, Guzzetti S, Rimoldi O, Furlan R, Pizzinelli P, Sandrone G, Malfatto G, Dell'Orto S, Piccaluga E et al (1986) Power spectral analysis of heart rate and arterial pressure variabilities as a marker of sympatho-vagal interaction in man and conscious dog. Circ Res 59(2):178–193

Paus T (2010) Population neuroscience: why and how. Hum Brain Mapp 31(6):891–903

Pausova Z, Abrahamowicz M, Mahboubi A, Syme C, Leonard GT, Perron M, Richer L, Veillette S, Gaudet D, Paus T (2010) Functional variation in the androgen-receptor gene is associated with visceral adiposity and blood pressure in male adolescents. Hypertension 55(3):706–714

Pausova Z, Paus T, Abrahamowicz M, Almerigi J, Arbour N, Bernard M, Gaudet D, Hanzalek P, Hamet P, Evans AC, Kramer M, Laberge L, Leal S, Leonard G, Lerner J, Lerner RM, Mathieu J, Perron M, Pike B, Pitiot A, Richer L, Seguin JR, Syme C, Toro R, Tremblay RE, Veillette S, Watkins K (2007) Genes, maternal smoking, and the offspring brain and body during adolescence: design of the Saguenay youth study. Hum Brain Mapp 28(6):502–518

Peltonen L, Palotie A, Lange K (2000) Use of population isolates for mapping complex traits. Nat Rev Genet 1(3):182–190

Schumann G, Loth E, Banaschewski T, Barbot A, Barker G, Büchel C, Conrod PJ, Dalley JW, Flor H, Gallinat J, Garavan H, Heinz A, Itterman B, Lathrop M, Mallik C, Mann K, Martinot JL, Paus T, Poline JB, Robbins TW, Rietschel M, Reed L, Smolka M, Spanagel R, Speiser C, Stephens DN, Ströhle A, Struve M (2010) IMAGEN consortium. The IMAGEN study: reinforcement-related behaviour in normal brain function and psychopathology. Mol Psychiatry 15(12):1128–1139

Stojiljkovic MP, Zhang D, Lopes HF, Lee CG, Goodfriend TL, Egan BM (2001) Hemodynamic effects of lipids in humans. Am J Physiol Regul Integr Comp Physiol 280(6):R1674–R1679

Syme C, Abrahamowicz M, Leonard GT, Perron M, Pitiot A, Qiu X, Richer L, Totman J, Veillette S, Xiao Y, Gaudet D, Paus T, Pausova Z (2008) Intra-abdominal adiposity and individual components of the metabolic syndrome in adolescence: sex differences and underlying mechanisms. Arch Pediatr Adolesc Med 162(5):453–461

Tanaka H, Thulesius O, Yamaguchi H, Mino M, Konishi K (1994) Continuous non-invasive finger blood pressure monitoring in children. Acta Paediatr 83(6):646–652

Taylor SE, Lerner JS, Sage RM, Lehman BJ, Seeman TE (2004) Early environment, emotions, responses to stress, and health. J Pers 72(6):1365–1393. doi:JOPY300 [pii] 10.1111/j.1467-6494.2004.00300.x

Tsui-Pierchala BA, Encinas M, Milbrandt J, Johnson EM, Jr. (2002) Lipid rafts in neuronal signaling and function. Trends Neurosci 25 (8):412–417. doi:S0166-2236(02)02215-4 [pii]

Vanhanen M, Koivisto K, Moilanen L, Helkala EL, Hanninen T, Soininen H, Kervinen K, Kesaniemi YA, Laakso M, Kuusisto J (2006) Association of metabolic syndrome with Alzheimer disease: a population-based study. Neurology 67(5):843–847. doi:10.1212/01.wnl.0000234037.91185.99

Votruba SB, Jensen MD (2007) Regional fat deposition as a factor in FFA metabolism. Annu Rev Nutr 27:149–163. doi:10.1146/annurev.nutr.27.061406.093754

Vyas AA, Schnaar RL (2001) Brain gangliosides: functional ligands for myelin stability and the control of nerve regeneration. Biochimie 83 (7):677–682. doi:S0300-9084(01)01308-6 [pii]

Waber DP, de Moor C, Forbes PW, Almli CR, Botteron KN, Leonard G, Paus T, Rumsey J, The Brain Development Cooperative Group (2007) The NIH MRI study of normal braindevelopment: performance of an epidemiologically ascertained sample of healthychildren aged 6 to 18 years on a neuropsychological battery. J Int Neuropsychol Soc 13:1–18

Chapter 9
Challenges

In science, the key challenge is to discover the truth. In population neuroscience, this means separating real relationships between exposures and outcomes from false ones. By design, cohort studies similar to those described in Chap. 8 generate massive amounts of data: thousands of system-level and molecular phenotypes, thousands of genotypes and hundreds of exposures related to our physical and social environments. When searching for "the truth" in such vast datasets, we can, in principle, use the following two strategies: (1) build multivariate statistical models to capture key relationships in the dataset and (2) approach the data in a quasi-experimental manner.

We will touch on the first approach in Chap. 10, when we discuss it in the context of risk/resilience profiling (Fig. 10.4). Here, we will outline a few strategies that are suitable for approaching the rich set of collected data as an experimental laboratory—where all necessary ingredients are ready for us to use in careful "bench work". In this context, we will talk about the use of meta-analyses for constructing hypotheses and discuss the importance of distinguishing between associations and causal relationships. We then reiterate the fact that MRI provides only a *representation* of the human brain: a few studies carried out in experimental animals will provide illustrations of how we can go beyond the "MR brain".

We close the chapter by pointing out a number of practical challenges, with a focus on incidental findings.

9.1 Constructing Hypotheses: Systematic Reviews and Meta-Analyses

Experiments start with a hypothesis. Most often, hypotheses are built on previous knowledge found in the published literature. Systematic reviews and meta-analyses represent tools for identifying consistent findings in a given area of inquiry and, as such, serve as an excellent starting point for constructing hypotheses.

In **epidemiology**, systematic reviews and meta-analyses allow one to synthesize all evidence available from observational studies and randomized trials. For

T. Paus, *Population Neuroscience*, DOI: 10.1007/978-3-642-36450-1_9,

example, the Cochrane Library (www.cochrane.org/cochrane-reviews/about-cochrane-library) contains systematic reviews in the area of health care, focusing on the results of randomized controlled trials (e.g. Johnston et al. 2012) and randomized clinical trials (e.g. Lee et al. 2011). In the case of observational studies, meta-analyses are particularly helpful for evaluating the consistency of the relationships between various exposures and outcomes, as observed across multiple studies, carried out in different settings and, often, with slightly different designs and measurement tools.

The main findings of such meta-analyses are typically presented as a forest plot of effect sizes, weighted by the variance and size of individual studies. Figure 9.1 shows a forest plot depicting the mean difference in cognitive-developmental score between breastfed and formula-fed children, adjusted for a number of possible confounders, including SES, maternal education, birth weight, parity, gestational age, maternal intelligence quotient and maternal smoking (Anderson et al. 1999). Such meta-analyses are invaluable in informing us about the expected effect size and potential confounders when studying the same exposure vis-à-vis brain structure and/or function (Isaac et al. 2010; Kafouri et al. 2012).

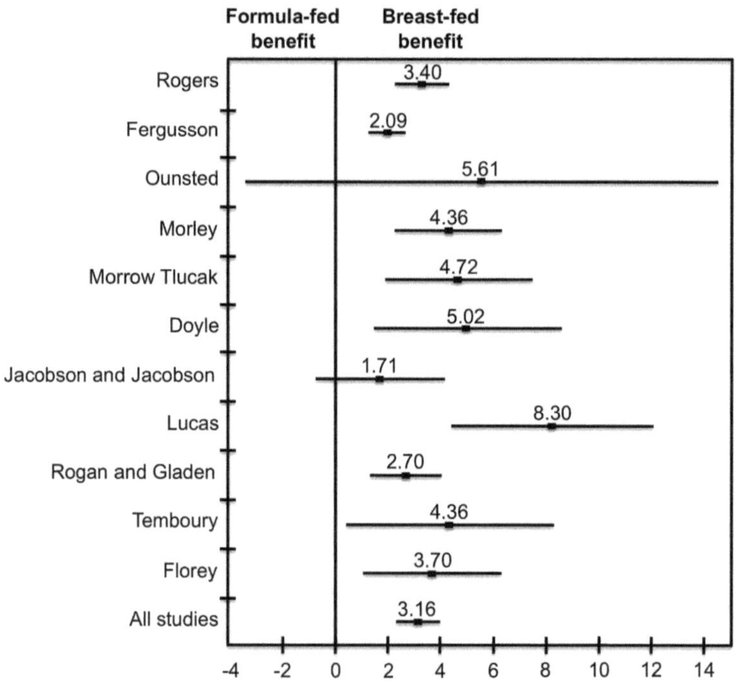

Fig. 9.1 Meta-analysis of the effects of breastfeeding on cognitive development. Numbers indicate co-variate-adjusted mean differences (and 95 % confidence intervals) for matched composite observations across 11 different studies. Reprinted from Anderson et al. (1999)

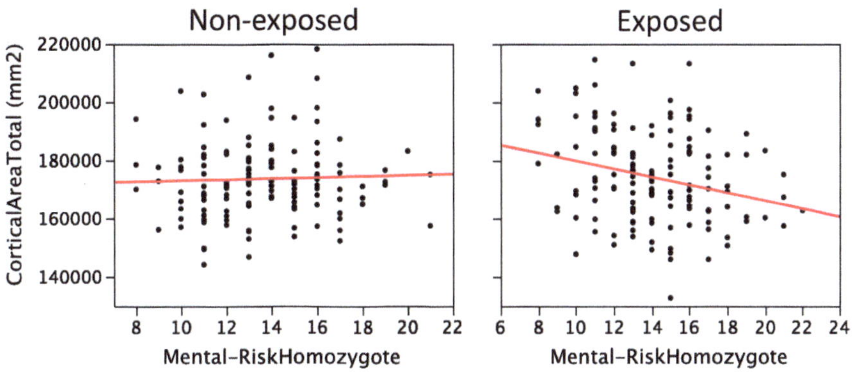

Fig. 9.2 Cortical area as a function of mutation load (number of homozygous genotypes) in exposed (R^2 = .07, p = .002) and non-exposed (R^2 = .00) adolescent females. From Bezgina, French, Paus (unpublished observation)

In **genetics**, we can use meta-analyses in order to narrow down the number of genes and/or genetic variants that can be tested for their associations with a given brain phenotype. Given the increasing availability of genome-wide coverage with SNPs, one way to achieve this is through meta-analyses of genome-wide association studies (see Sect. 4.5). Using the catalogue of published GWAS (Hindorff et al. 2012; www.genome.gov/gwastudies/), one can select, for example, a set of SNPs associated—across multiple studies—with a particular group of psychiatric disorders. A somewhat different approach would target all mutations, such as single-point non-synonymous SNPs or small deletions/insertions, linked to a particular disorder through linkage-based analyses (see Sect. 4.5). The human gene mutation database (HGMD, www.hgmd.cf.ac.uk/ac/index.php; see Sect. 4.4) has catalogued over 100,000 disease-causing and disease-associated germ-line DNA mutations, affecting ~3,700 human genes (Cooper et al. 2010). We have used this resource to identify 300+ non-synonymous SNPs contained in our genotyping platform (Illumina HumanOmniExpress BeadChip) and—based on HGMD—linked with various psychiatric disorders, such as autism, schizophrenia, bipolar disorder and Alzheimer's disease. We then asked whether there is a cumulative effect for being a risk-allele homozygote across these 300+ mutations on the overall cortical surface. As shown in Fig. 9.2, this was the case, but only for female adolescents exposed prenatally to maternal cigarette smoking (Bezgina, French, Paus, unpublished observation). Thus, in addition to the candidate gene and candidate panel approaches for selecting relevant genetic variants, meta-analyses of the catalogued gene–phenotype relationships provide a nice entry point for brain-based analyses.

Finally, can we use meta-analyses for targeting the most relevant **brain phenotype**? One approach available to us is that of the "activation likelihood estimation" (ALE; Eickhoff et al. 2009), a meta-analytical approach allowing one to identify brain regions that show consistent response to a given stimulus

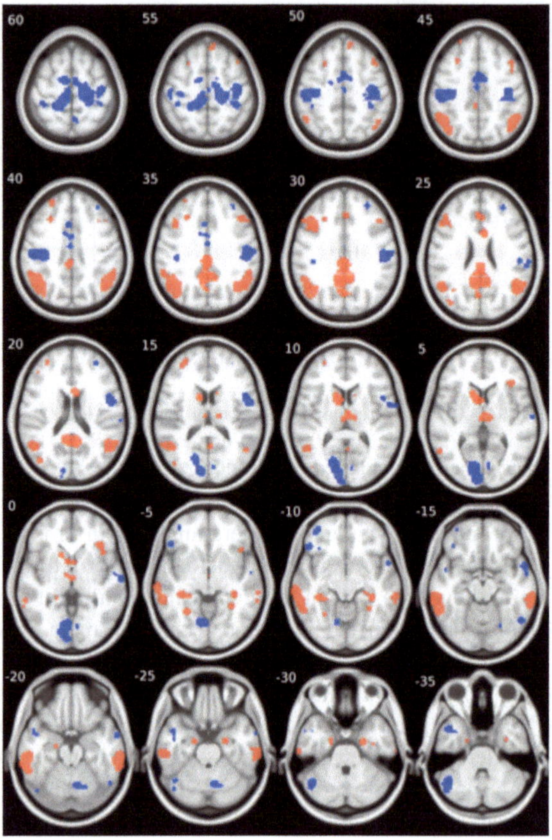

Fig. 9.3 "Vulnerable" (*red*) and "resistant" (*blue*) cortical regions. The "vulnerable" regions were identified as showing consistent hypometabolism in resting fluorodeoxyglucose scans in patients with Alzheimer's disease, as compared to age-matched controls. The "resistant" (control) regions were chosen by positioning an equivalent set of regions in parts of the brain not containing the "vulnerable" regions. Numbers indicate Z coordinates of the axial slices in the ICBM152 stereotaxic space. From Schwartz and Paus (unpublished observation)

(e.g. a face) across multiple imaging studies.[1] The ALE approach can be applied to any imaging modality, functional (e.g. fMRI, PET) or structural (e.g. DTI, VBM), as long as the results have been based on a voxel-wise analysis carried throughout the entire brain and reported using a standardized coordinate system. For example, we have constructed an ALE map of group differences in glucose metabolism—between 915 patients with Alzheimer's disease and 715 age-matched controls, scanned in 33 studies (Fig. 9.3; Schwartz and Paus, unpublished observation). In this way, we have identified a "vulnerable" and "resistant" brain space in which to test effects of various exposures and genes.

[1] One can also use the ALE approach to identify a consistent difference, across multiple studies, between two groups of individuals (e.g. patients with Alzheimer's disease and healthy controls).

9.2 Association Versus Causality

With observational studies, we tend to identify correlations ("breastfeeding is associated with intelligence") rather than cause-and-effect relationships ("breastfeeding affects intelligence"). Of course, the most desirable findings are those that indicate a causal relationship between exposure and outcome. Is there some way of injecting causality into observational studies?

The use of genetics through the so-called **Mendelian randomization** is one way of doing so. Mendelian randomization is a special case of the instrumental-variable approach used in econometrics and epidemiology to infer causality (Bowden and Turkington 1984; Greenland 2000; Hernan and Robins 2006). Here, a functional genetic variant (an instrumental variable), which is "randomized" with regards to the inheritance of other traits, acts as a proxy for an environmental exposure without being confounded by other variables (Ebrahim and Davey Smith 2008; Sheehan et al. 2008). For example, the known association between breastfeeding and IQ may be due to a number of factors—from mother–infant attachment to various macro- and micronutrients contained in breast milk. A study of a gene variant (FADS2) known to influence plasma levels of polyunsaturated fatty acids suggests that, at least in part, this particular exposure "leads" to the higher IQ (Caspi et al. 2007). In our work, we used this approach to show that the hormonal environment, namely testosterone, "affects" the volume of white matter during male adolescence. Male adolescents with an "efficient" (short) version of the androgen receptor (*AR*) gene showed much stronger relationship between testosterone and the volume of white matter than males with less "efficient" (long) *AR* (Fig. 9.4; Perrin et al. 2008). This suggests that the androgenic effect of testosterone indeed "causes" the increase in the volume (see below, for additional follow-ups of this observation).

Causality and its direction can also be determined by **temporal criteria** (i.e. the cause must precede the effect). Provided enough data have been acquired with an appropriate sampling frequency, the directionality of "transactions" occurring across time between hypothetical exposures and outcomes can be examined with appropriate statistical models (e.g. time-lagged analysis; see Sect. 1.5 on developmental cascades). This is very important given the bidirectional nature of brain–behaviour relationships (see Sect. 7.1). Let us assume, for example, that we assessed repeatedly a large sample of typically developing adolescents for the use of marijuana, the presence of psychosis-like symptoms and their brain structure, at 12, 16 and 20 years of age. As illustrated in Fig. 9.5, the difference in the size of the nucleus accumbens at 12 years of age would predict the use of marijuana at 16 years of age, and the use of marijuana at 16 years of age would predict, in turn, white matter properties and the presence of psychosis-like symptoms at 20 years of age.

This sequence of events suggests that unknown prepubertal factors (genetic and/or environmental) underlie the inter-individual variation in the size of the nucleus accumbens and that, in turn, increases the probability of smoking marijuana at 16. Marijuana use then leads to the simultaneous emergence of changes in

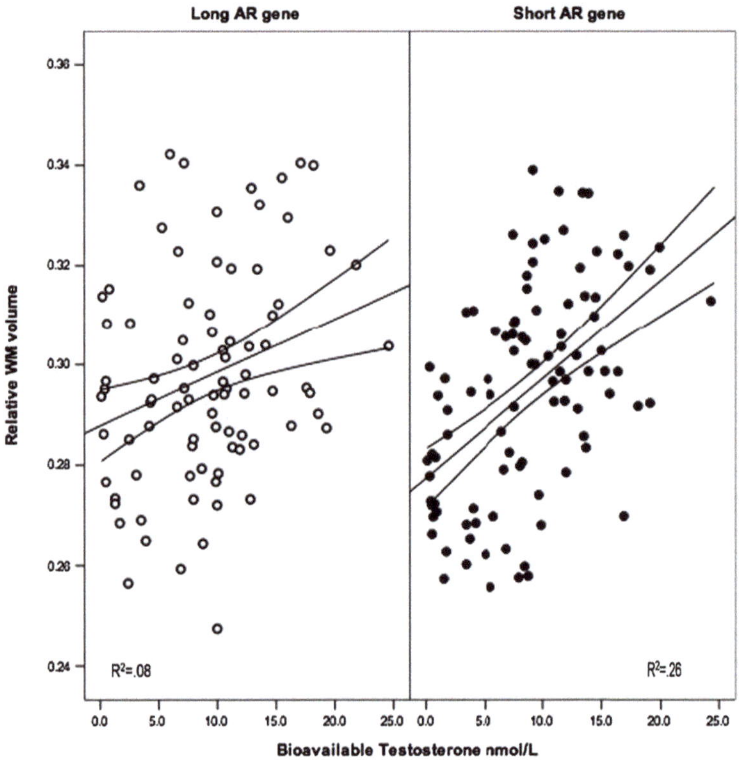

Fig. 9.4 Relative volume of white matter (WM) plotted as a function of plasma levels of bioavailable testosterone in male adolescents with high (long AR gene) and low (short AR gene) number of CAG repeats in the AR gene. From Perrin et al. (2008)

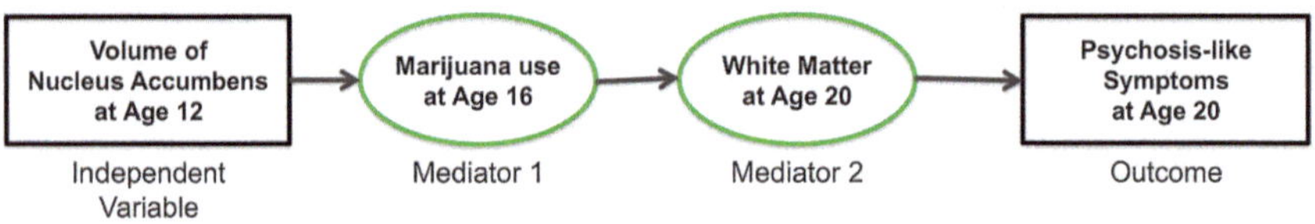

Fig. 9.5 Cascade of hypothetical events leading to the occurrence of changes in white matter and psychosis-like symptoms at 20 years of age

the white matter and psychosis-like symptoms by the time the individual reaches 20 years of age. Needless to say, this timeline alone is insufficient for establishing an unequivocal cause-and-effect relationship between the above events. If combined with the Mendelian randomization (e.g. functional polymorphisms in the cannabinoid receptor *CB1* gene), it would represent an excellent basis for an intervention designed as a randomized controlled trial.

Randomized controlled trials represent the ultimate test of causality available in human populations. Exposures are assigned at random in a sample drawn from the general population. Relevant phenotypes are measured before and after the exposure. But for a number of reasons—including ethics, feasibility and cost—few exposures can be tested in this manner. Yet it is possible to design relevant interventions using ethically acceptable alternatives. For example, one cannot assign randomly selected newborns to be either breastfed or formula fed. As implemented in the promotion of breastfeeding intervention trial (PROBIT), however, it was possible to randomize individual maternity hospitals (in a country) into ones that comply with breastfeeding practices based on the WHO/UNICEF Baby-Friendly Hospital Initiative vs. those that continue with the practices in place at the time of randomization (Kramer et al. 2002). Using this approach, breastfeeding was found to be associated with higher verbal IQ but not with variations in BMI or blood pressure in childhood (Martens 2012). Overall, as suggested by Wareham et al. (2008), "Mixed approaches combining data from large-scale observational studies with smaller intervention trials may be ideal".

Of course, **experimental studies** carried out in vivo (animal models) or in vitro (e.g. brain slices, tissue cultures) provide an excellent vehicle for testing causality of the relationships observed in human studies. In the Saguenay Youth Study, for example, we have shown that prenatal exposure to maternal cigarette smoking is associated with drug experimentation (i.e. number of substances tried in the lifetime) during adolescence (Lotfipour et al. 2009). Even though we controlled for a number of potential confounders, we could not be certain about the causality of this relationship. Also, given that cigarette smoke contains thousands of chemicals, we could not simply attribute this "effect" to nicotine. This is why a study carried out by Franke et al. (2008) in rats was essential for guiding the interpretation of our findings. In this study, adolescent rats prenatally exposed to nicotine showed no inclination to self-administer cocaine at a low dose (Fig. 9.6a); they did so, however, when offered a high dose (Fig. 9.6b). The opposite was true about the non-exposed rats. By using an experimental manipulation, this finding reveals a causal relationship between prenatal exposure to nicotine and adolescent behaviour. It also offers a possible interpretation of our human studies that nicotine lowered the reward threshold, possibly leading to further search for a substance associated with (a stronger) reward. This is but one illustration of the importance of combining both experimental and population neurosciences.

9.3 Beyond the MR Brain

Without the doubt, MRI is a powerful technique for deriving a large number of structural and functional brain phenotypes. But when interpreting these MR-based measures, we must keep in mind that MR provides but a *representation* of the human brain—a representation built on specific acquisition sequences and image analysis methods (as reviewed in Chap. 7). It is therefore critical that we keep checking the veracity of such representations, the "MR brain", against the neurobiological reality. Let us illustrate this challenge with two examples drawn

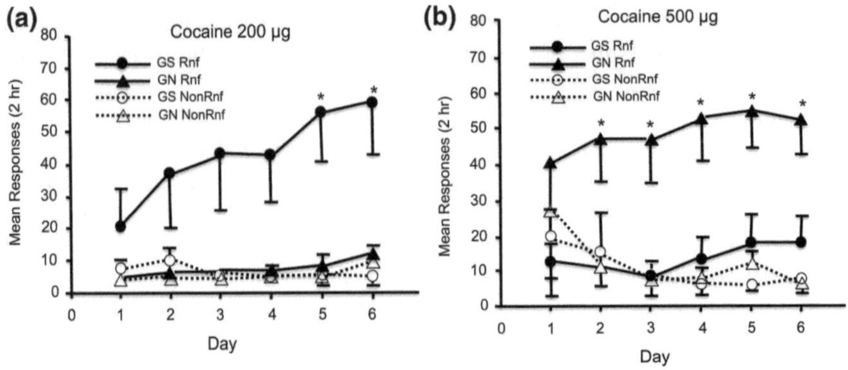

Fig. 9.6 Self-administration of cocaine by rats exposed (*triangles*) and not-exposed (*circles*) prenatally to nicotine. *GS*, gestational saline; *GN*, gestational nicotine; Rnf, reinforced; Non-Rnf, not reinforced. From Franke et al. (2008)

from the work on two structural features relevant to the study of white matter, namely a fibre path and a fibre diameter.

Mapping **a fibre path** is the key objective of DTI-based tractography (see Sect. 7.3.5). Using diffusion-weighted images, various algorithms are used to derive a representation of a fibre tract (Chung et al. 2011). How close are such representations to reality? Seehaus et al. (2012) have asked this question using a *post-mortem* sample of (human) temporal lobe (Text Box 9.1.).

Text Box 9.1. Tractography: Does a representation fit the reality?

Here, the fibre path representation was obtained from diffusion-weighted images (obtained at 9.4 T) with a line propagation algorithm, while the true fibre path was based on the diffusion of fluorescent lipophilic tracers (DiI and DiA) implanted just under the cortical surface. Given the slow diffusion of these tracers, they could examine only fairly short paths (from 3 to 9 mm). At an optimal value of fractional anisotropy, DTI-based tracts that were initiated (seeded) in a voxel containing the dye ("true pathways") were more likely to pass through other voxels containing the same dye, as compared to the (control) tracts seeded in voxels not containing the dye ("false pathways"; Figs. 9.7). As can be seen in Fig. 9.8b, however, the match is not perfect: some of the tractography-based representations of fibres originating outside the (*red*) dyed fibres nevertheless overlap with these fibres.
Note that the closeness of the match between the representation and the reality is the highest over very short distances (~3 mm), as compared to the medium (~5 mm) and "far" (~8 mm) distances. It is quite likely that the veracity of tractography-based representations will be much lower for images obtained at a lower field (e.g. at the common 3 T, as compared to 9.4 T used here), over longer distances (~50–100 mm), and in the living (moving and breathing) individual.

Fig. 9.7 Histogram of numbers of true voxels passed by true tractography pathways (*light grey*) and false tractography pathways (*dark grey*). Many more true voxels are passed by true pathways than by false pathways. From Seehaus et al. (2012)

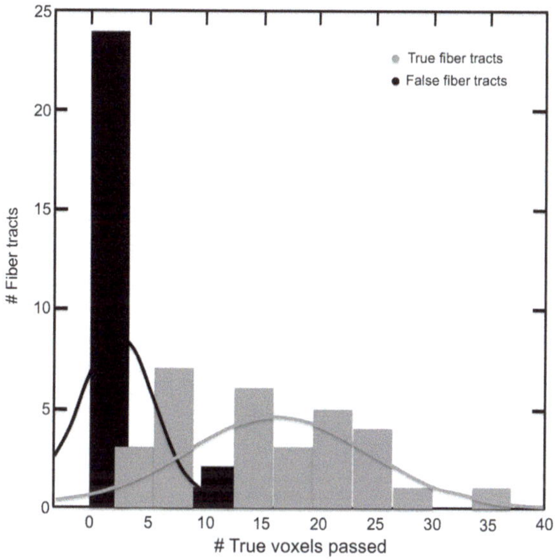

The volume of white matter is determined by the number of fibres, glial cells and the **fibre diameter**. In our work on the adolescent brain, we found that the volume of WM increases more steeply with age in males than in females (Perrin et al. 2008; Fig. 9.4). We then asked whether this volumetric increase is driven by an increase in myelination (i.e. increase in the thickness of the myelin sheath) or an increase in axon diameter. In the same sample, we were able to use MTR values (in WM) as an index of myelination. To our surprise, MTR values *decreased* with age in male adolescents (Fig. 9.9).

Rather than interpreting this finding as a decrease in myelination, we have suggested that the apparent decrease in myelination is, in fact, due to an increase in axon diameter; thicker axons "dilute" the myelin-related signal in a unit of scanned volume (Paus and Toro 2009; see also Herve et al. 2009, 2011). In

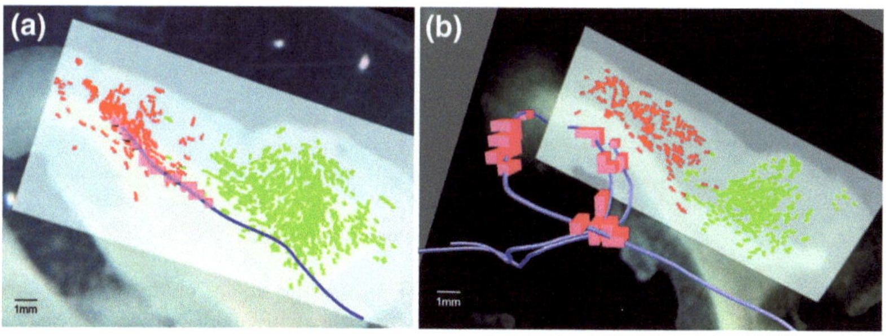

Fig. 9.8 (**a**) True-positive tractography pathway passing through many true voxels (*transparent red*). *Red* and *green*: different fluorescent crystal tracings. (**b**) False fibres that nevertheless contain many true voxels. From Seehaus et al. (2012)

addition, thicker axons have a relatively thinner myelin sheath (e.g. Chatzopoulou et al. 2008), thus decreasing the myelin-to-axon ratio in the scanned volume. Of course, these were all only inferences based on indirect evidence. We needed to check the reality with a tool that allows us to "see" the actual axon and the myelin sheath: electron microscopy. For this reason, we have measured axon diameter (Fig. 9.10a) and *g* ratio (the ratio between axon diameter and fibre diameter; Fig. 9.10b) in the corpus callosum of male and female rats.

These two examples illustrate the importance of checking the neurobiological "reality" behind the MR-based representations. In the first case (fibre path), this was achieved by a simultaneous application of MR and histology in a sample of the human brain. In the second example (fibre diameter), we used a histological technique alone and applied it to test a sex difference (in the rat brain) predicted by a previous MR study (in humans).

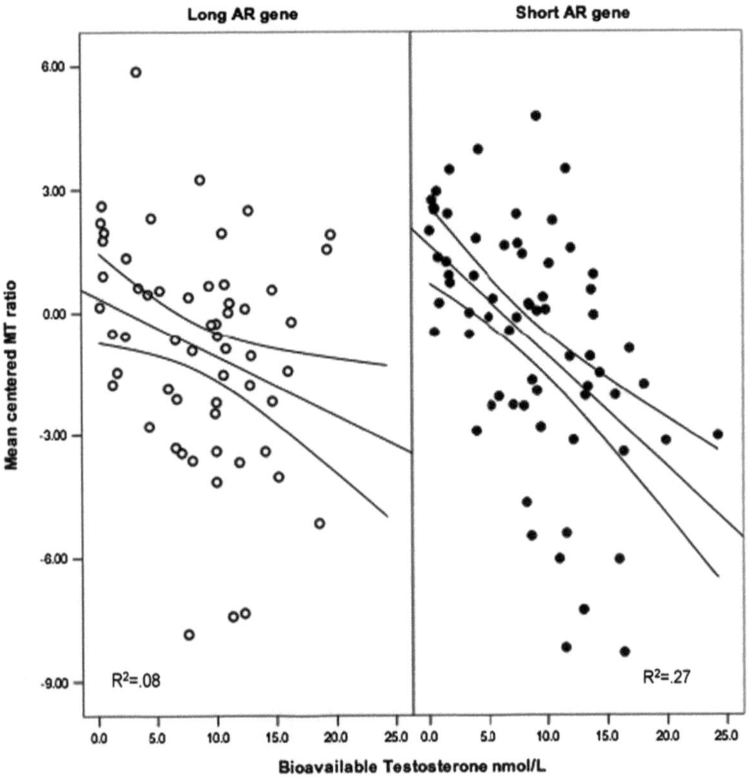

Fig. 9.9 Values of magnetization transfer ratio (MTR) in white matter plotted as a function of plasma levels of bioavailable testosterone in male adolescents with high (long AR gene) and low (short AR gene) number of CAG repeats in the AR gene. From Perrin et al. (2008)

Overall, there are, of course, multiple challenges associated with population-based studies similar to those described in Chap. 8. Here, we touched on a few theoretical challenges while leaving aside the practical ones. And there are many of those: ascertainment of participants (see Text Box 8.1.); their retention for longitudinal studies; quality assurance and quality control of MR imaging; data transfer and data basing; issues associated with the protection of confidentiality; and reporting of the incidental findings. Many of these challenges have been covered previously (Paus and Toro 2009). Before closing this chapter, we will address briefly one issue that comes up very often when designing large-scale MR studies: **incidental findings**.

In Canada, Article 3.4 of the Tri-council Policy Statement on Ethical Conduct for Research Involving Humans (2010) states: "Researchers have an obligation to disclose to the participant any material incidental findings discovered in the course of research". This article defines "incidental findings" as "unanticipated discoveries, made in the course of research but that are outside the scope of the research". It states further that "material incidental findings are findings that have been interpreted as having significant welfare implications for the participant, whether health-related, psychological or social". Most experts in medical ethics and law agree on the following criteria for the "return" of findings: their analytical

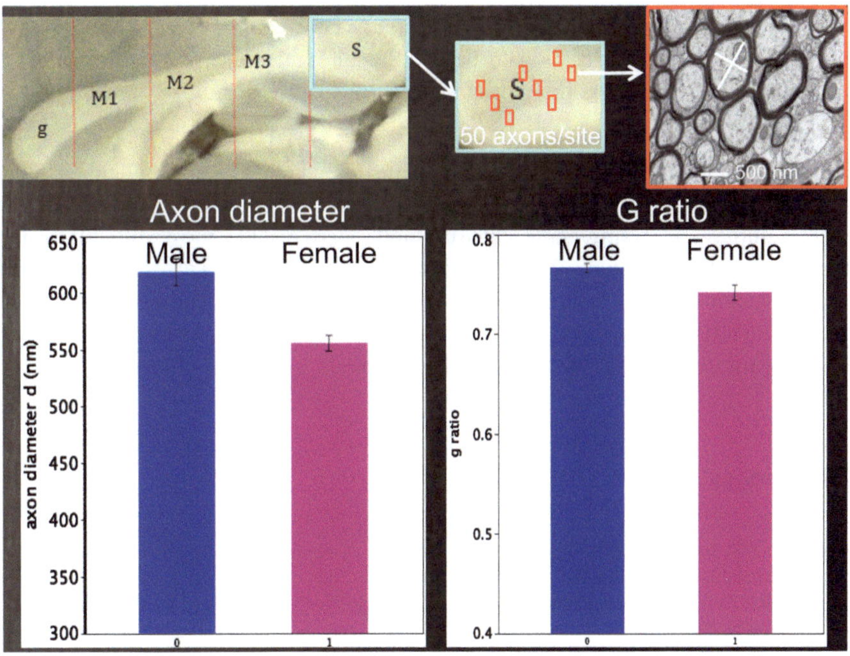

Fig. 9.10 Sex differences in axon diameter and *g* ratio in the rat corpus callosum (splenium, S). *G* ratio is the ratio between axon diameter and fibre diameter, the latter being the sum of the axon diameter and myelin thickness. Higher *g* ratio reflects relatively thinner myelin sheath in thicker axons. From Pesaresi and Paus (unpublished observations)

validity, clinical utility and "actionability"[2] (e.g. Bredenoord et al. 2011). We have been guided by these principles in our previous and ongoing MR-based population studies, carried out in different jurisdictions, including the Canadian provinces of Ontario and Quebec, as well as in the United States and several European countries (www.brainbody.org/studies.html; Text Box 9.2.).

Text Box 9.2. Procedures for Reporting Incidental Findings

Our policy begins by informing the participant during the recruitment and consent process that "[t]he proposed MR scan is research specific and of minimal medical diagnostic value. A radiologist will **not** examine the images on a routine basis". We then inform the participant that "should a researcher involved in the project incidentally observe a suspicious finding, the scan will be shown to a radiologist. If the radiologist considers the finding to be of medical significance, [the participant] and/or [his/her] doctor will be contacted". In practice, MR technologists are most often the ones who observe incidental findings at the time of the MR exam. In the event of this happening, they file an "incidental finding" report and forward it to the study coordinator. He/she contacts a radiologist to request a radiological review of the scans. For a fee, a radiologist will review each incidental finding, make a recommendation as to its clinical utility (and actionability) and return the recommendation to the study coordinator. If the radiologist deems the incidental finding to be of clinical significance, the coordinator will communicate it to the participant and/or his/her doctor.

In closing, practitioners of population neuroscience face many challenges in the road to discovery: from designing their studies, collecting multi-modal data in large samples, analysing these datasets and—perhaps the most difficult one—interpreting their findings. We all want to find the right answer—and an answer that matters in terms of both the new knowledge and its possible impact on human health. Effective team work, scientific rigour and personal perseverance are the essential ingredients for success in this enterprise.

References

Anderson JW, Johnstone BM, Remley DT (1999) Breast-feeding and cognitive development: a meta-analysis. Am J Clin Nutr 70(4):525–535

Bowden RJTD (1984) Instrumental variables. Cambridge University Press, Cambridge

[2] The term "actionability" is used by ethicists and lawyers to indicate whether or not "something can be done" with regards to the individual's health (e.g. a surgery to remove a tumour).

Bredenoord AL, Kroes HY, Cuppen E, Parker M, van Delden JJ (2011) Disclosure of individual genetic data to research participants: the debate reconsidered. Trends Genet 27(2):41–47. doi:S0168-9525(10)00227-1 [pii] 10.1016/j.tig.2010.11.004

Caspi A, Williams B, Kim-Cohen J, Craig IW, Milne BJ, Poulton R, Schalkwyk LC, Taylor A, Werts H, Moffitt TE (2007) Moderation of breastfeeding effects on the IQ by genetic variation in fatty acid metabolism. Proc Natl Acad Sci U S A 104(47):18860–18865. doi:0704292104 [pii] 10.1073/pnas.0704292104

Chatzopoulou E, Miguez A, Savvaki M, Levasseur G, Muzerelle A, Muriel MP, Goureau O, Watanabe K, Goutebroze L, Gaspar P, Zalc B, Karagogeos D, Thomas JL (2008) Structural requirement of TAG-1 for retinal ganglion cell axons and myelin in the mouse optic nerve. J Neurosci 28(30):7624–7636

Chung HW, Chou MC, Chen CY (2011) Principles and limitations of computational algorithms in clinical diffusion tensor MR tractography. AJNR Am J Neuroradiol 32(1):3–13. doi:ajnr.A2041 [pii] 10.3174/ajnr.A2041

Cooper DN, Chen JM, Ball EV, Howells K, Mort M, Phillips AD, Chuzhanova N, Krawczak M, Kehrer-Sawatzki H, Stenson PD (2010) Genes, mutations, and human inherited disease at the dawn of the age of personalized genomics. Hum Mutat 31(6):631–655. doi:10.1002/humu.21260

Ebrahim S, Davey Smith G (2008) Mendelian randomization: can genetic epidemiology help redress the failures of observational epidemiology? Hum Genet 123(1):15–33. doi:10.1007/s00439-007-0448-6

Eickhoff SB, Laird AR, Grefkes C, Wang LE, Zilles K, Fox PT (2009) Coordinate-based activation likelihood estimation meta-analysis of neuroimaging data: a random-effects approach based on empirical estimates of spatial uncertainty. Hum Brain Mapp 30(9):2907–2926. doi:10.1002/hbm.20718

Franke RM, Park M, Belluzzi JD, Leslie FM (2008) Prenatal nicotine exposure changes natural and drug-induced reinforcement in adolescent male rats. Eur J Neurosci 27(11):2952–2961. doi: EJN6253 [pii]10.1111/j.1460-9568.2008.06253.x

Greenland S (2000) An introduction to instrumental variables for epidemiologists. Int J Epidemiol 29(6):1102

Hernan MA, Robins JM (2006) Instruments for causal inference: an epidemiologist's dream? Epidemiology 17(4):360–372. doi:10.1097/01.ede.0000222409.00878.37

Herve PY, Cox EF, Lotfipour AK, Mougin OE, Bowtell RW, Gowland PA, Paus T (2011) Structural properties of the corticospinal tract in the human brain: a magnetic resonance imaging study at 7 Tesla. Brain Struct Funct 216(3):255–262. doi:10.1007/s00429-011-0306-0

Herve PY, Leonard G, Perron M, Pike B, Pitiot A, Richer L, Veillette S, Pausova Z, Paus T (2009) Handedness, motor skills and maturation of the corticospinal tract in the adolescent brain. Hum Brain Mapp 30(10):3151–3162

Hindorff LA, MacArthur J (European Bioinformatics Institute), Wise A, Junkins HA, Hall PN, Klemm AK, Manolio TA (2012) A Catalog of Published Genome-Wide Association Studies. Available at: www.genome.gov/gwastudies

Isaacs EB, Fischl BR, Quinn BT, Chong WK, Gadian DG, Lucas A (2010) Impact of breast milk on intelligence quotient, brain size, and white matter development. Pediatr Res 67(4):357–362. doi:10.1203/PDR.0b013e3181d026da

Johnston V, Liberato S, Thomas D (2012) Incentives for preventing smoking in children and adolescents. Cochrane Database Syst Rev 10:CD008645. doi:10.1002/14651858.CD008645.pub2

Kafouri S, Kramer M, Leonard G, Perron M, Pike B, Richer L, Toro R, Veillette S, Pausova Z, Paus T (2012) Breastfeeding and brain structure in adolescence. Int J Epidemiol. doi:10.1093/ije/dys172 dys172 [pii]

Kramer MS, Guo T, Platt RW, Shapiro S, Collet JP, Chalmers B, Hodnett E, Sevkovskaya Z, Dzikovich I, Vanilovich I (2002) Breastfeeding and infant growth: biology or bias? Pediatrics 110(2 Pt 1):343–347

Lee MS, Kim JI, Ernst E (2011) Massage therapy for children with autism spectrum disorders: a systematic review. J Clin Psychiatry 72(3):406–411. doi:10.4088/JCP.09r05848whi

Lotfipour S, Ferguson E, Leonard G, Perron M, Pike B, Richer L, Seguin JR, Toro R, Veillette S, Pausova Z, Paus T (2009) Orbitofrontal cortex and drug use during adolescence: role of prenatal exposure to maternal smoking and BDNF genotype. Arch Gen Psychiatry 66(11):1244–1252

Martens PJ (2012) What do Kramer's baby-friendly hospital initiative PROBIT studies tell us? A review of a decade of research. J Hum Lact 28(3):335–342. doi:10.1177/0890334412438264 0890334412438264 [pii]

Paus T, Toro R (2009) Could sex differences in white matter be explained by g ratio? Front Neuroanat 3:14

Perrin JS, Herve PY, Leonard G, Perron M, Pike GB, Pitiot A, Richer L, Veillette S, Pausova Z, Paus T (2008) Growth of white matter in the adolescent brain: role of testosterone and androgen receptor. J Neurosci 28(38):9519–9524

Seehaus AK, Roebroeck A, Chiry O, Kim DS, Ronen I, Bratzke H, Goebel R, Galuske RA (2012) Histological validation of DW-MRI tractography in human post-mortem tissue. Cereb Cortex. doi:10.1093/cercor/bhs036 bhs036 [pii]

Sheehan NA, Didelez V, Burton PR, Tobin MD (2008) Mendelian randomisation and causal inference in observational epidemiology. PLoS Med 5(8):e177. doi:10.1371/journal.pmed.0050177 08-PLME-RIT-0499 [pii]

Tri-Council Policy Statement: Ethical Conduct for Research Involving Humans (2010) Canadian Institutes of Health Research, Natural Sciences and Engineering Research Council of Canada, and Social Sciences and Humanities Research Council of Canada, December 2010

Wareham NJ, Young EH, Loos RJ (2008) Epidemiological study designs to investigate gene-behaviour interactions in the context of human obesity. Obesity (Silver Spring) 16(Suppl 3):S66–S71. doi:10.1038/oby.2008.521 oby2008521 [pii]

Chapter 10
Personalized Preventive Medicine

As described in the previous chapters, population neuroscience brings together tools and concepts enabling us to identify which elements in our genes and environment are crucial for shaping the human brain, from conception onwards. Ultimately, we strive to understand and clarify all important factors and mechanistic pathways associated with the incredible diversity of brain phenotypes. Of course, such knowledge has one practical application: the ability to predict the risk of brain disorders and, in turn, use those predictions to prevent or mitigate the suffering involved.

In this chapter, we explore the possibility of predicting and preventing common disorders of brain *and* body, thus highlighting the importance of the integrated approach introduced in Chap. 8.

10.1 Healthy Life Expectancy: Common Chronic Disorders of Brain and Body

Healthy life expectancy is equivalent to the number of years lived in full health. Across the 192 WHO Member States, the gap between (total) life expectancy (life span) and healthy life expectancy (health span) is 7.5 years (11.5 % of total life expectancy). The number of lost healthy years ranges from 9 % in Europe to 15 % in Africa (Mathers et al. 2004). These numbers are based on life tables, disease prevalence and self-reports from the patient indicating health decrement. Although the "health span" metrics may vary somewhat with type of source data and adjustments made, they are particularly useful in developed countries with stable social systems. For example, data based on the annual Health Survey for England (1994–1999) show a clear relationship between (socio-economic) deprivation scores and healthy life expectancy (Fig. 10.1). The number of *unhealthy* years varies from 11 years (14.5 % of total life expectancy) to 22 years (31 %) for males living in the least and most deprived areas, respectively (Bajekal 2005).

Given the demographic shifts occurring in developed countries, increasing *healthy life expectancy* is one of the critical elements of their healthcare systems.

T. Paus, *Population Neuroscience*, DOI: 10.1007/978-3-642-36450-1_10,

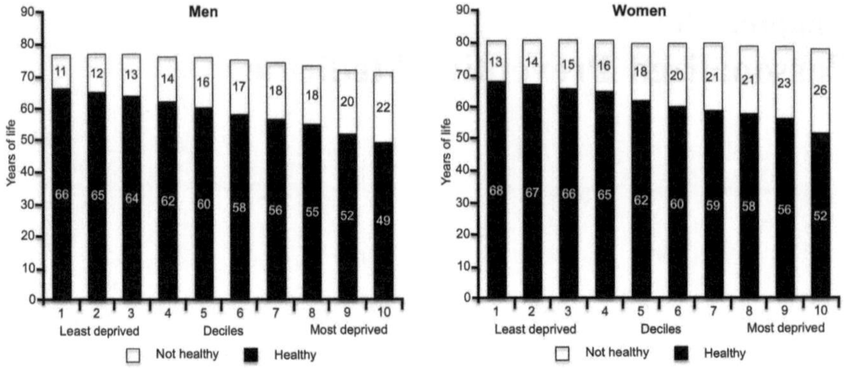

Fig. 10.1 Healthy and total (healthy + unhealthy) life expectancy at birth in England, 1994–1999. Adapted from the Office for National Statistics licensed under the Open Government Licence v.1.0. From Bajekal (2005)

Healthy life expectancy provides a strong impetus for change, from a "reactive" to a "proactive" kind of medicine (see below). While twentieth century medicine was about finding causes and treatments, twenty-first century medicine looks for mechanistic pathways and preventive strategies.

What are the most common conditions associated with lower healthy life expectancy? In developed countries, the leading causes of disability—expressed as years lived with a disability—are neuropsychiatric conditions (41.9 %), cardiovascular conditions (6.7 %), respiratory (6.9 %) and musculoskeletal (7.6 %) diseases, malignant neoplasms (2.4 %) and diabetes mellitus (2.3 %) (Mathers et al. 2004). Among neuropsychiatric conditions, the most significant causes of disability are depression (15 %), alcohol-use disorders (6.8 %) and Alzheimer's and other dementias (4.2 %) (Mathers et al. 2004). The number of years lived with disability (YLD) is rising, not only due to the increase in total life expectancy but also due to the increasing prevalence of diseases associated with a sedentary lifestyle (e.g. Kohl et al. 2012; Tremblay et al. 2011); moderate levels of physical activity (brisk walking daily) can increase life expectancy by as much as 4 years (Moore et al. 2012). Many of the major causes of disabilities listed above fall into this category.

Let us now consider the following common disorders of the brain and body: depression, addictions, dementias and what is known as metabolic syndrome (high blood pressure, impaired fasting glucose or insulin resistance, dyslipidemia, central obesity). The total number of YLD due to these disorders accounts for as much as 38.9 % of all YLD from all causes (Mathers et al. 2004). This is due to their prevalence (Table 10.1), tendency to emerge early in life (adolescence) and chronic course. Furthermore, living with these disorders is associated with a significant loss of the quality of life. This can be estimated by calculating preference scores for a given health state, which are typically expressed on a scale ranging

Table 10.1 Prevalence, the number of healthy life years lost in the Americas due to a given condition and the annual cost of common chronic disorders in Canada

Disease	Prevalence			Health years lost—Americas	Annual cost (CDN)—Canada
	Adolescents (12–19 years)	Middle aged (20–64 years)	Aged (65+ years)		
Depression	6.4	5.0	1.9	10,764,000	$1.4 billion
Addiction					$39.8 billion
Alcohol	7.0	2.0	Unavailable	4,830,000	$14.6 billion
Illegal drugs	2.7	0.5	Unavailable	2,433,000	$8.2 billion
Dementia	Unavailable	Unavailable	8.0	2,209,000	$15 billion
Obesity	19.7	18.0	19.1	Unavailable	$4.3 billion
Hypertension	0.6	12.8	48.9	1,105,000	$880 million
Diabetes	Unavailable	4.2	18.1	4,095,000	$5.6 billion
Stroke	Unavailable	1.3	8.5	3,988,000	$3.6 billion
Acute myocardial infarction	Unavailable	1.7	10.2	6,523,000	$3.0 billion
Cancer	0.06	1.0	6.2	11,461,000	$14.2 billion

Numbers for cancer are provided for comparison

from 1 (ideal health) to 0 (equivalent to being dead). For example, in a sample drawn from general population, living with moderate or severe depression is associated with preference scores of 0.7 and 0.5, respectively (Pyne et al. 2009). Not surprisingly, the economic cost of living with these disorders is staggering (Table 10.1). Depression alone ranks among major causes of disability worldwide—the second largest in industrialized countries (Lopez et al. 2006). In Canada, one estimate of the annual cost of depression, addictions and dementias combined is $50 billion.

Furthermore, as we pointed out in Sect. 8.3.2, common disorders of the body, such as obesity, diabetes, hypertension and dyslipidemia, are associated with impaired brain health. Overall, increasing with age, there is significant **co-morbidity** across the brain and body disorders (Fig. 10.2). Thus, common psychiatric co-morbidities of depression are anxiety disorders and substance (alcohol) abuse (e.g. Schuckit 2006; Krishnan 2005). Some consider depression to be one of the risk factors for dementia (Tsuno and Homma 2009). Among medical co-morbidities of depression, the most prevalent and/or clinically relevant are cardiovascular disease, obesity and diabetes. The mortality rate for patients with depression is twice that of the general population and is mainly due to increased cardiovascular mortality and suicides (Osby et al. 2001). Finally, dementia is more than twice as likely to occur in patients with metabolic syndrome (Vanhanen et al. 2006; see Sect. 8.3.2). This underlines the importance of an integrated brain–body approach when searching for risk profiles predicting brain health. Table 10.2 illustrates this point in terms of co-morbidities of bipolar disorders (Krishnan 2005).

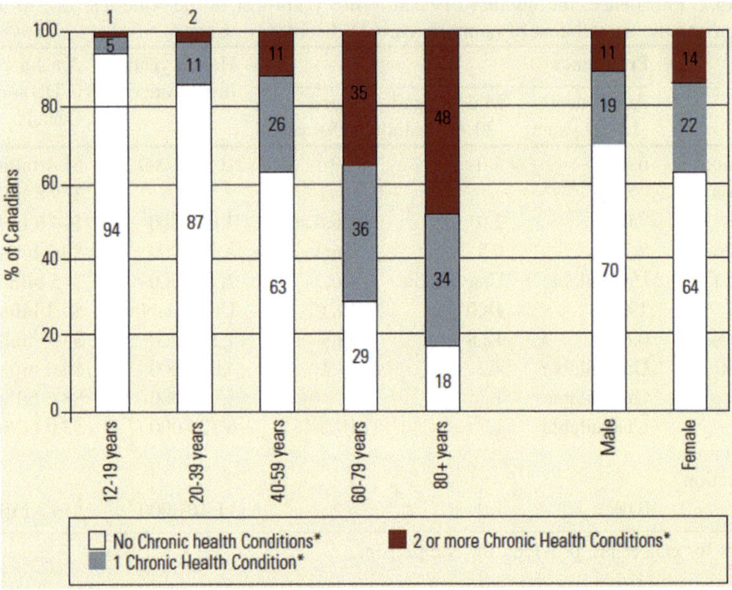

Fig. 10.2 Percentage of Canadians with none, one or two and more chronic health conditions (by age). *Select chronic health conditions include arthritic, cancer, chronic obstructive pulmonary disease, diabetes, heart disease, high blood pressure and mood disorders. Source: Statistics Canada (2005). From Broemeling et al. (2008).

Table 10.2 Lifetime co-morbidities of bipolar disorder with other disorders. *PTSD* post-traumatic stress disorder; *OCD* obsessive–compulsive disorder

Co-morbid condition	Mean rate of co-morbidity (%)	Percentage range across studies (%)
Any axis I disorder	65	50–70
Substance use disorder	56	34–60
Alcohol abuse	49	30–69
Other drug abuse	44	14–60
Anxiety disorder	71	49–92
Social phobia	47	
PSD	39	
Panic disorder	11	3–21
OCD	10	2–21
Binge-eating disorder	13	
Personality disorder	36	29–38
Migraine	28	15–40
Overweight	58	
Obesity	21	
Type 2 diabetes	10	
Hypothyroidism	9	

From Krishnan (2005)

10.2 Personalized Preventive Medicine

One size does not fit all. In the general population, the clinical manifestations (symptoms) and pathophysiology of the chronic disorders mentioned above are far from uniform. To start with, the prevalence rates of most of these disorders vary by sex (e.g. higher female-to-male ratio for depression [1.69] and Alzheimer's dementia [1.99] and lower for alcohol-use disorders [0.24]; Mathers et al. 2004), age and ethnicity (Chiu et al. 2010). Pathways leading to these disorders are also likely to differ across different segments of the general population. For example, Chiu and colleagues reported a relatively low incidence of heart disease in Black Canadians, despite the high composite index of relevant risk (Chiu et al. 2010). This might indicate an uncoupling somewhere along the risk pathways or the existence of a resilience mechanism that counteracts the detrimental effects of these risk factors.

Another example of such an atypical profile is the virtual absence of obesity-associated hypertension in Pima Indians in Arizona, despite the presence of the typical association between type 2 diabetes and obesity in the same group (Hanson et al. 2002). In this case, the uncoupling may occur between fat cells (e.g. production of free fatty acids [FFA] and/or FFA-related signalling) and the autonomic nervous system. Persistent activation of the ANS is one pathophysiological mechanism leading to hypertension (Pausova 2006). Can detailed knowledge of an individual's brain and body, gained through the approaches described here, be translated into ways of improving his/her health?

10.2.1 Personalized Medicine

Personalized medicine has emerged as a new paradigm that attempts to tailor treatment strategies to a specific patient (Hamburg and Collins 2010). In this context, biological markers (or biomarkers) are defined as any of an individual's measured characteristics that may strongly indicate the suitability of a particular (drug) treatment for that individual (Anonymous 2001). Typically, biomarkers index the state of key pathogenic mechanisms underlying a given disease or an individual's response to a specific treatment. But in the context of making long-term predictions, we need (preclinical) biomarkers *unaffected* by the full-blown diseases and its treatment (Text Box 10.1).

Text Box 10.1. Preclinical biomarkers

Biomarkers can index the state of key mechanistic pathways leading to a given disease. This can be achieved at different levels: (1) *genomic level* (e.g. a DNA variation associated with the low activity of an enzyme); (2) *transcriptomic level* (e.g. expression level [RNA] of this enzyme in the relevant tissue); (3) *proteomic level* (e.g. levels of this enzyme in blood or the relevant tissue); and (4) *systems level* (e.g. state of the organ/system impacted by the action of this enzyme, such as the structural and functional properties of vessels). Note that studying young- and middle-aged individuals in a preclinical stage of disorders is essential for identifying biomarkers (and mechanistic pathways) free of confounders associated with a full-blown (clinical stage) disease and its treatment.

At present, pharmacogenetics is perhaps the most promising example of personalized medicine. For example, the Food and Drug Administration in the United States approved several diagnostic tests to predict the effectiveness of specific anti-cancer drugs, based on the expression of particular genes in the tumour (Hambug and Collins 2010). Similarly, a dosage of an anti-coagulant drug (warfarin) is influenced by variants of the genes encoding specific metabolizing enzymes (*CYP2C9*, *VKORC1*), among European Americans but not African Americans (Limdi et al. 2008).

These examples illustrate the promise as well as the challenges of personalized medicine. A single biomarker is likely to have limited predictive value for an individual, especially when applied in the context of a complex disorder. For this reason, attention is now shifting towards "biomarker signatures/patterns" or "profiles of change" (Barker 2004; Scherer et al. 2010), again emphasizing the importance of high-dimensional and high-throughput phenomics. This topic is addressed below in the section on the risk/resilience profiling.

10.2.2 Preventive Medicine

Preventive medicine is likely to benefit from the same personalized approach as described for personalized medicine above. One of the challenges of implementing preventive medicine lies in the effectiveness of global public health initiatives aimed at reducing known risk factors for chronic disorders, such as metabolic syndrome. Despite clear benefits, a large segment of the general population fails to follow *broad* (i.e. generic) health recommendations. For example, the EPIC-Norfolk study, carried out in the United Kingdom between 1993 and 2006, revealed that only 21 % of men and 39 % of women engaged in all four of the healthy behaviours studied: being physically active, eating enough fruits and vegetables, avoiding excessive use of alcohol and not smoking (Khaw et al. 2008). These four factors carried a benefit of a fourfold reduction in the relative risk of mortality from all causes (including a fivefold reduction in cardiovascular mortality risks), compared with those who did not engage in those healthy behaviours at all. This is equivalent to adding 14 years of chronological age to an individual's life expectancy (Khaw et al. 2008).

As a way of complementing global health promotion strategies (those applied to the general population), the healthcare system is beginning to focus its prevention strategies on individuals at risk of developing a disease (primary prevention) or on stopping a disease (or co-morbidity) in its early stages (secondary prevention). The main challenge of this approach is to identify at-risk individuals. As discussed extensively by the British epidemiologist Geoffrey Rose, a single biomarker is likely to identify individuals at risk of a disease only when it reaches extreme values; thus, only individuals at *very high* risk are identified (Rose 2008). But, by definition, this happens only in a small fraction of the general population. In the majority of individuals, the biomarker has as an intermediate value—making it difficult to decide with confidence whether or not a given individual is at risk of the disease in question. In terms of public health, this poses a conundrum (Text Box 10.2).

Text Box 10.2. Estimating risk using a single biomarker

Let us have a look at an example of glycated haemoglobin (HbA1C), a marker of glucose levels in blood in past months (and, in turn, diabetic state and its compensation), and its association with mortality (from Rose 2008). Rose notes that intermediate levels of HbA1C are associated with a relatively low relative risk (RR) of mortality (RR = 1.5), as compared to its high levels (RR = 3.5). On the other hand, because of the higher prevalence of intermediate ($\sim$24 %) versus high ($\sim$4 %) values in the general population, the estimated excess mortality rate associated with intermediate values is greater for intermediate versus high values of HbA1C ($\sim$33 vs. $\sim$24 %, respectively) (Rose 2008). Thus, we cannot really say whether or not an individual with an intermediate level of HbA1C would become sick and die earlier, and yet, at a population level, these individuals contribute more to the overall mortality attributable to this phenotype than individuals with high levels.

As pointed out by Khaw and Marmot in their commentary on Rose's Strategy of Preventive Medicine, "the majority of cases in the population occur not in the small numbers at very high risk but in the centre of the population distribution, where large numbers of people are exposed, albeit with only modest increases in risk" (Rose 2008, p. 7). Thus, one solution for **personalized preventive medicine** is to identify (the large number of) individuals at modest risk of developing a disease and offer them personalized strategies for risk reduction. But this cannot be achieved with a single biomarker. We suggest, instead, the development of risk/resilience profiling, based on multiple characteristics measured objectively in each individual, with state-of-the-art technologies operating at both a molecular and a system level. Clearly, one hopes that if we combine such high-fidelity information about the individual's phenome with details about his/her genome and envirome, we will attain a level of accuracy to justify actions tailored to a particular individual.

10.3 Risk/Resilience Profiling

Profiling has been used in medicine from the very beginning. It is the specific combination of symptoms that leads to a correct diagnosis and any available treatment (McInnes and Semple 1994). For example, using data collected in the Framingham Heart Study, a 10-year risk of coronary heart disease (myocardial infarction and coronary death) can be calculated based on a number of (now) known risk factors, including age, total cholesterol, HDL cholesterol, systolic blood pressure, treatment for hypertension and cigarette smoking (Framingham Risk Scores; Anonymous 2002, 2010). Note, however, that this index fails to identify a great proportion of future coronary heart disease (Thompson et al. 2006).

A profile may indicate an optimal combination of physiological parameters associated with, for example, a long-term history of success in running a marathon (Maud et al. 1981). In general, the predictive value of a profile reflects the current state of pertinent knowledge. Conceptually, a profile serves a dual purpose: (1) a practical one (i.e. as a predictor) and (2) a theoretical one (i.e. as a model of the current state of knowledge). As explained below, phenotypic rather than genetic profiling may provide more power for building such predictive models. Nonetheless, we will begin by reviewing current concepts of genetic profiling—and their critiques—before turning our attention to the phenotypic profiling approach.

10.3.1 Genetic Profiling

Genetic profiling emerged as a possible tool for predicting an individual's risk of complex diseases, using variations in his or her genome. Unlike monogenetic disorders—for example, Huntington's disease (Fig. 10.3a) or phenylketonuria (Fig. 10.3b)—complex diseases are likely caused by host of factors: multiple genes (G1–G4 in Fig. 10.3c), various environmental influences (Fig. 10.3e) and, of course, combinations of the two (Fig. 10.3d, f).

Thus, "complex diseases result from the joint effects of multiple genetic and environmental causes, with each factor having only minor contribution to the occurrence of the disease" (Janssens and van Duijn 2008). Not surprisingly, we gain very little information by predicting the risk of such complex disorders using a single genetic variant (Vineis et al. 2001). This explains the shift towards simultaneous testing at multiple genetic loci, in other words, genetic profiling. But, as reviewed by Janssens and van Duijn (2008), genetic profiling for complex disorders has not been successful thus far; discriminative accuracy varies between 0.55 and 0.8 (0.5 = chance; 1.0 = perfect discrimination).

Even more important is the observation that genetic profiling adds very little, if anything, to the predictive value of clinical risk factors associated with complex disorders—including known biomarkers such as body mass index for type 2 diabetes (Janssens and van Duijn 2008). As concluded by these authors, "genetic

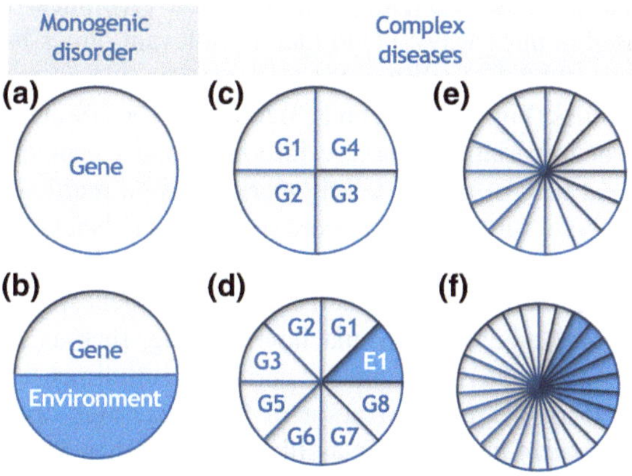

Fig. 10.3 Complete cause models or sufficient causes of disease development, G, gene; E, environment. From Janssens and van Duijn (2008)

testing may not improve the prediction of disease beyond classical risk factors or new biomarkers, if most of the genes involved in the disease play a role through these risk factors". On the other hand, they underscore the heuristic value of genetic profiling for revealing new pathophysiological pathways underlying these disorders and, in turn, new opportunities for the development of novel interventions and prevention strategies (Janssens and van Duijn 2008). It should be pointed out that the above discussion is limited to profiling based on genetic (DNA) variations in the genome: **genomic profiling** may be a better term to use here. At present, it is unknown whether variations in the epigenome (**epigenomic profiling**) or in gene expression (**transcriptomic profiling**) may add unique predictive value, especially if assessed in relevant tissues.

10.3.2 Phenotypic Profiling

Phenotypic profiling—the measurement of the set of functional and structural characteristics at a specific point in an individual's life—has the advantage of capturing the cumulative effects of both genetic and environmental factors.

One can generate risk/resilience profiles using a combination of multi-level (e.g. molecules, systems) and high-dimensional (e.g. genome, epigenome, phenome, envirome) approaches. In this way, we could encapsulate a large number of quantitative characteristics relevant to the common, chronic disorders of the brain and body. We would do so by using MRI, mental health and cognitive assessments, cardiometabolic assessments, transcriptomics and proteomics, as well as an assessment of relevant social and physical environments. Building multivariate statistical models to capture key relationships in the dataset complements the quasi-experimental approached described in Chap. 9.

As shown in Fig. 10.4, the modelling work begins with the derivation of phenotypic risk profiles (Modelling 1), followed by the generation of genotypic risk profiles (Modelling 2). Note that preclinical (system-level) phenotypes are used in three ways: (1) to identify relevant omics-based profiles (Fig. 10.4-A); (2) to identify novel genetic risk profiles via phenotype-based multidimensional profiles (Fig. 10.4-B); and (3) to identify a subset of any known disease-associated genetic variations that are informative at the preclinical level (Fig. 10.4-C). The great bulk of the modelling work relies on multivariate techniques, such as factor analysis, multivariate z-score transforms (Mahalanobis' distance) and independent component analysis (ICA), the latter used in parallel mode when searching for profiles across two domains (e.g. brain–genes; Meda et al. 2012). One can also employ methods from machine learning, such as support vector machines, which are particularly suitable for deriving multidimensional profiles given their ability to learn the categorization of complex, high-dimensional training data and to generalize the learned classification rules then apply them to new data (e.g. Hastie et al. 2001).

Using such an approach, one can attempt to capture, for example, the relevant interactions between the brain and body systems leading to the development of dementia (Text Box 10.3 and Fig. 10.5). The phenotypic profiles are complemented by an environmental profile derived from questionnaires and sensor-based recordings of lifestyle (yellow), plus genomic, epigenomic and transcriptomic profiles of relevant molecules. The latter would help uncover moderating and mediating influences of specific molecular pathways.

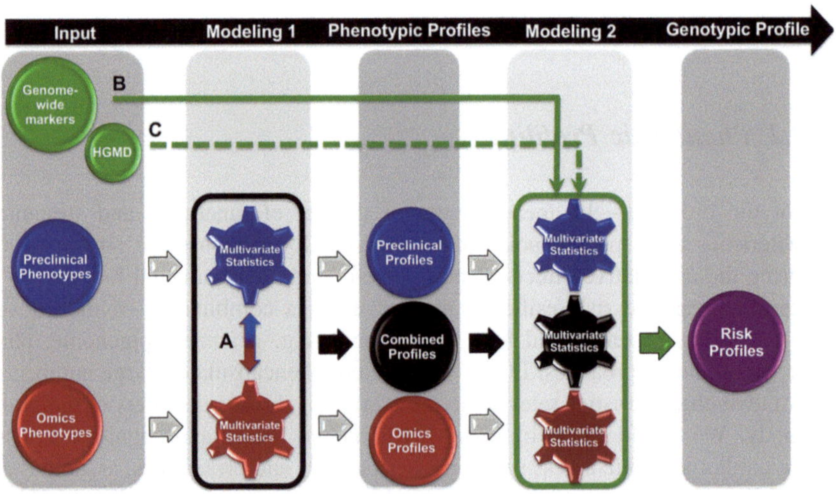

Fig. 10.4 A multi-step identification of risk profiles. Preclinical phenotypes are based on multimodal system-level assessments of brain and behaviour. Omics phenotypes may include epigenomics, transcriptomics, metabolomics and lipidomics. *HGMD* the Human Gene Mutation Database

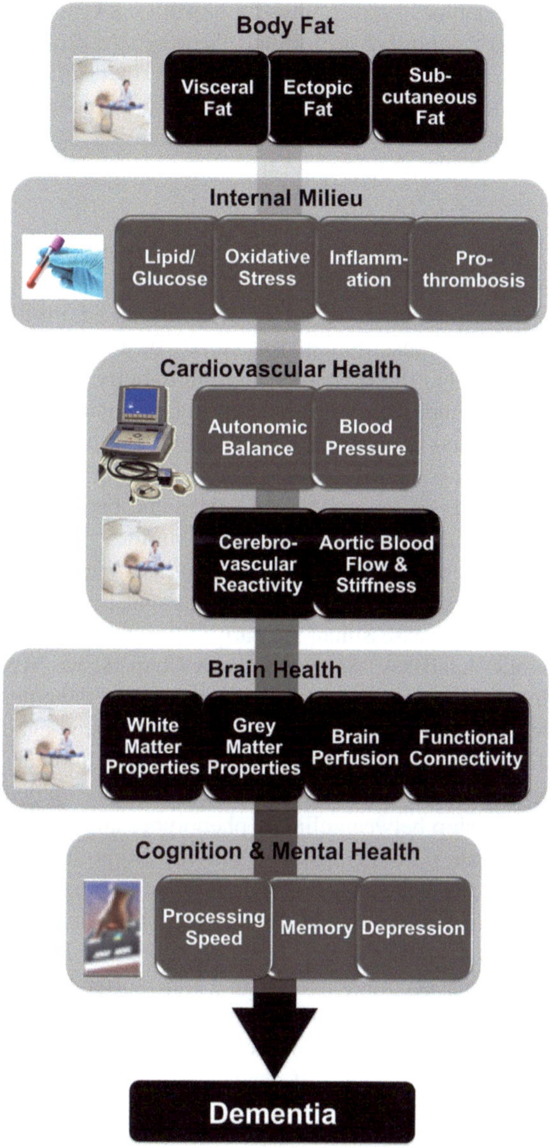

Fig. 10.5 Key metabolic, cardiovascular and brain phenotypes underlying a hypothetical metabolic–cognitive profile that predicts the probability of developing dementia. MRI-based phenotypes are purple. Measures of sympathoactivation and blood pressure are derived from beat-by-beat recordings obtained with a Finometer. The assessment of lipids, glucose and insulin and the inflammatory milieu is carried out through lipidomic and proteomic assays. See Lusis et al. (2008) for a similar scheme capturing the pathophysiology of heart failure due to metabolic syndrome

Text Box 10.3. Stages of Phenotypic Profiling

Phenotypic profiling may proceed in stages, from building a single-phenotype profile (e.g. the Finometer-based time series of systolic blood pressure or the MRI-based time series of cerebrovascular reactivity [CVR]), through organ-based profiles (e.g. a composite brain profile built on a combination of phenotypes constituting, for example, a CVR profile, a white matter profile and a cognitive profile), all the way to a cross-organ (fitness) profile, combining phenotypes fed into the brain profile, cardiovascular profile (e.g. Finometer time series, cardiac output, aortic stiffness), metabolic profile (e.g. lipid profile, visceral and ectopic fat, glucose and insulin) and inflammatory profile (cytokines and other inflammatory biomarkers).

How can we validate the predictive value of such profiles? One possible strategy would be to take advantage of a trans-generational design (Sect. 8.3.1) and test the predictive value of a family-based profile (Fig. 8.3.). Thus, one can evaluate the discriminative accuracy of a "Prognosis by Ancestor/Pedigree".

Given (1) the relatively high heritability of the many quantitative phenotypes likely to underlie a profile based, for example, on the characteristics (and relevant biomarkers) depicted in Fig. 10.5 and (2) the number of common environmental factors (e.g. energy excess) shared during certain critical periods of life (e.g. childhood) by the members of the same family, it is likely that a number of key pathogenic processes will follow similar virtual trajectories among the related members of three-generational families. We can also consider inter-generational differences related to secular trends and immigration, captured by the questionnaire-based assessment of early environmental exposures and through relevant public databases. As described in Chap. 8, we hypothesize that the accuracy of discriminating between a descendant (e.g. daughter or granddaughter) who will develop a disease and one who will not, tested against the profile (and/or disease status) of her ancestor (e.g. mother or grandmother), will be comparable with the discriminative accuracy observed in recent epidemiological studies, where the relationship between clinical phenotypes and disease status is evaluated on a short time scale ($\sim$0.80; Janssens and van Duijn 2008).

In a nutshell, the prognostic accuracy of the profiles is tested against the health status (e.g. the presence or absence of a clinical diagnosis) of the grandparents (virtual trajectory). But, ultimately, one needs to validate the profile by testing prognostic accuracy against real trajectories in a prospective manner; at a minimum, one can track an individual's health status using clinical health databases that provide information about hospitalizations, use of prescription medications and, ultimately, cause of death.

10.4 My Health Trajectory

As pointed out above, there is a pressing need for shifting the healthcare paradigm from reactive to preventive medicine. As shown in Table 10.3, we have at our disposal a wide variety of tools to prevent, delay or mitigate common disorders of the brain and body.

Table 10.3 Matching predictors and interventions

Prediction	Intervention
Molecules	Pharmacological
Systems	Pharmacological
Individuals	Health behaviours
Built environments	Building design, architecture
Organizations (schools, workplace, food supermarkets)	Local policies
Neighbourhood	Urban planning
Community	Community centres and programs, health clinics
Society	Healthcare system, education, taxation

Here, we focus on discussing a possible way for the knowledge gained through the risk/resilience profiling to be brought directly to "customers", enabling them to take proactive action regarding their health. This approach, **personalized preventive medicine**, (1) targets individuals; (2) empowers them to make decisions at multiple levels (e.g. management of personalized disease risk, dietary risk, lifestyle risk); (3) brings personalized health science to both lay clients and front-line health professionals (e.g. family doctors, nutritionists, lifestyle coaches); and (4) has a strong personal appeal.

We have begun conceptual work for developing a tool for such a client-driven personalized preventive medicine, namely the My Health Trajectory portal. This portal would provide personalized profiles that are informative with regards to clients' short- and long-term risk of common chronic disorders of the brain and body. In this way, My Health Trajectory would inform individuals and their care team (e.g. family physicians) about their unique predispositions. This would go beyond family history shared, for example, by two brothers. It would also help them track the health consequences of improvements (or declines) in their behaviour and/or of implementing other preventive interventions (including treatment-based secondary interventions). Ultimately, we and others believe that personalized information is the first step towards a truly person-based system of health promotion.

There are a few precedents for this "direct-to-consumer" approach to providing genomic information, albeit currently limited. Several companies now offer personal genome services via Web-based portals, such as 23andMe (www.23andme.com) and deCODEme (www.decodeme.com). It is of note that, in July 2012, 23andMe filed for US Food and Drug Administration clearance for 7 of its 200-plus genetics tests (see Allison 2012). These "direct-to-consumer" services typically include the following three steps: (1) mailing a DNA collection kit to the client; (2) extracting and genotyping DNA with a set of markers (e.g. 23andMe uses the Illumina OmniExpress Plus chip); and (3) providing this information to the client and/or his/her physician through a secure website. Some companies also provide clients with access to board-certified genetic councillors. The three most common domains targeted by these companies include the following: (1) determining a carrier status for monogenic disorders; (2) estimating a client's response to a number of drugs; and (3) estimating a client's risk of a number of complex (polygenic) diseases ad traits (Text Box 10.4.). In all cases, these estimates are based on published literature, contained in various databases and evaluated by company staff and advisors.

Text Box 10.4. Direct-to-Consumer genomics

A few examples of personalized genomic information provided by 23andMe. (1) A carrier status for monogenic disorders: 44 diseases, such as *BRCA* or familial hypercholesterolaemia; (2) A response to a number of drugs: 20 drugs, such as warfarin or statins; (3) A risk of a number of complex (polygenic) diseases: 116 disease, such as alcohol dependence or hypertension; and (4) A risk of a number of complex traits: 55 traits, such as food preference or HDL/LDL cholesterol levels.

While My Health Trajectory would build on this basic approach, it would introduce several innovations. First of all, estimates of genetic risk for the common chronic disorders would be based primarily on genetic profiles derived through the trans-generational approach described above and, as such, should have a stronger predictive value than single predictors derived from published data. Second, risk profiles would extend beyond the static genes into the dynamic (i.e. modifiable) epigenome, lipidome and metabolome. In this manner, one would be able to estimate short-term risks often associated with specific lifestyle-induced environments (as reflected in the "omics" data). Importantly, the modifiable nature of these profiles would provide clients with a yardstick to measure the impact of behavioural changes and/or treatment-based interventions carried out in the context of secondary prevention (e.g. statins vs. neuroprotective agents in individuals with metabolic syndrome to prevent/delay onset of dementia) they would make in response to receiving their first profile. Third, the system-level (MRI-based) phenotyping of the brain and body would allow one to evaluate the added value of such imaging-based profiling; if judged useful, one could develop an abbreviated version of the scanning protocols to be incorporated as a product in My Health Trajectory.

As shown in Fig. 10.6, the My Health Trajectory portal would be designed as a layered ("choose-a-product") system for delivering personalized information about (1) genetic and epigenetic profiles; (2) current lipidomic and metabolomic profiles; (3) the current state of the body; and (4) the brain, the latter two assessed with MRI.

For a number of reasons, the time is right for this kind of approach: (1) high-tech health science (e.g. omics and imaging) can be achieved with high throughput at a reasonable cost; (2) the public—especially the young- and middle-aged target population—is technologically savvy; and (3) both young and old are taking their

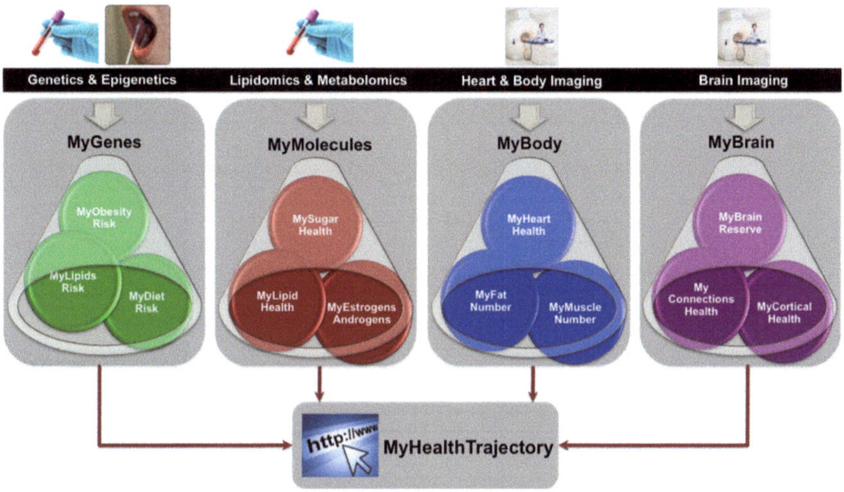

Fig. 10.6 A proposed My Health Trajectory portal

health into their own hands in growing numbers. What we need to do is to develop tools that would increase their *motivation* to make the right health choices.

10.5 A Case Study: Delaying Onset of Alzheimer's Disease

Let us conclude this chapter by considering whether or not personalized preventive medicine could delay the onset of Alzheimer's disease, a major public health concern in the ageing populations of today's world. Achieving this goal would make a huge difference to the social and economic health of societies facing the challenge of dealing with the cognitive decline of their ageing populations.

As estimated in the 2010 report produced by the Alzheimer's Association, delaying the onset of Alzheimer's dementia by 5 years would result in a decrease in the projected number of patients with this disease from 13.5 to 7.7 million (United States, year 2050; Fig. 10.7a). This translates into a savings of US $447 billion (Anonymous 2010; Fig. 10.7b).

But what kind of intervention could delay the onset by 5 years? The authors of the above report state that "The hypothetical treatment might be a vaccine that would be given once in a person's life, a medication or cocktail of medications that would be taken one or more times a day, starting at different times in a person's life, or a change in diet, exercise or other lifestyle behaviors" (Anonymous 2010, p. 5).

Given our emphasis on personalized prevention, both primary and secondary, let us examine *modifiable* risk factors for Alzheimer's disease. As shown in Table 10.4, these factors—diabetes, midlife hypertension, midlife obesity, smoking, depression, cognitive inactivity or low educational attainment, and physical

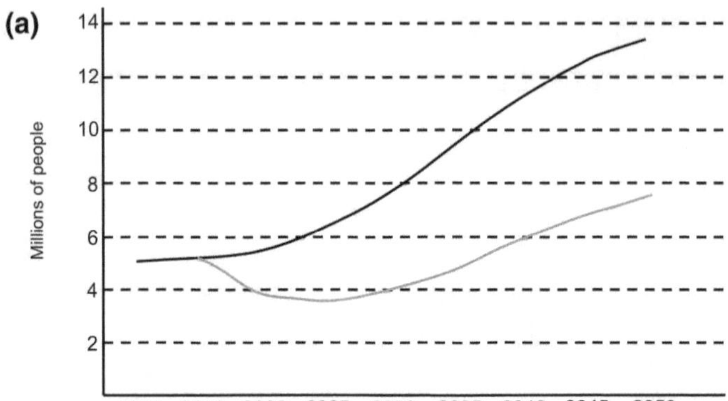

	2010	2015	2020	2025	2030	2035	2040	2045	2050
Current Trajectory	5.1	5.3	5.6	6.5	7.8	9.5	11.2	12.7	13.5
Delayed Onset	5.1	5.3	4.0	3.8	4.2	5.1	6.1	7.1	7.7
Decrease		0.0	1.6	2.7	3.6	4.4	5.1	5.6	5.8

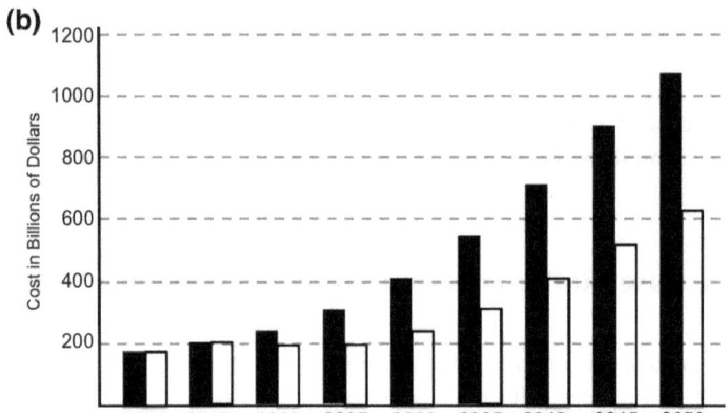

	2010	2015	2020	2025	2030	2035	2040	2045	2050
Current Trajectory	$172	$202	$240	$307	$408	$547	$717	$906	$1078
Delayed Onset	$172	$202	$190	$196	$239	$311	$407	$522	$631
Reduced Cost	$0	$0	$50	$111	$170	$236	$310	$384	$447

Fig. 10.7 **a** Americans aged 65 and older with Alzheimer's disease, 2010–2015. **b** Impact of a 5 year delay in onset on costs (in US dollars); Americans aged 65 and older with Alzheimer's disease, 2010–2015. From Anonymous (2010)

inactivity—contribute an incredible 50 % to the population-attributable risk of Alzheimer's disease (Barnes and Yaffe 2011).

Not surprisingly, a 10–25 % reduction in all seven *modifiable* risk factors would lead to a significant drop in the number of patients (Barnes and Yaffe 2011; Fig. 10.8).

How do we modify—that is, decrease—these risk factors, in a given individual rather than on a population level? Here, we come back to the two basic tenets of personalized preventive medicine: (1) one size does not fit all and (2) knowledge is power. The first idea simply reiterates the fact that pathogenic pathways leading to complex disorders, such as Alzheimer's disease, are not identical across individuals.

Table 10.4 Cases of Alzheimer's disease attributable to potentially modifiable risk factors worldwide and in the USA

	Population prevalence (%)	Relative risk (95 % CI)	Population-attributable risk (confidence range)	Number of case attributable (thousands; confidence range)
Worldwide				
Diabetes mellitus	6.4	1.39 (1.17–1.66)	2.4 % (1.1–4.1)	826 (365–1,374)
Midlife hypertension	8.9	1.61 (1.16–2.24)	5.1 % (1.4–9.9)	1,746 (476–3,369)
Midlife obesity	3.4	1.60 (1.34–1.92)	2.0 % (1.1–3.0)	678 (387–1,028)
Depression	13.2	1.90 (1.55–2.33)	10.6 % (6.8–14.9)	3,600 (2,295–5,063)
Physical inactivity	17.7	1.82 (1.19–2.78)	12.7 % (3.3–24.0)	4,297 (1,103–8,122)
Smoking	27.4	1.59 (1.15–2.20)	13.9 % (3.9–24.7)	4,718 (1,338–8,388)
Low education	40.0	1.59 (1.35–1.86)	19.1 % (12.3–25.6)	6,473 (4,163–8,677)
Combined (maximum)			50.7 %	17 187 028[a]
USA				
Diabetes mellitus	8.7	1.39 (1.17–1.66)	3.3 % (1.5–5.4)	174 (77–288)
Midlife hypertension	14.3	1.61 (1.16–2.24)	8.0 % (2.2–15.1)	425 (119–789)
Midlife obesity	13.1	1.60 (1.34–1.92)	7.3 % (4.3–10.8)	386 (226–570)
Depression	19.2	1.90 (1.55–2.33)	14.7 % (9.6–20.3)	781 (506–1078)
Physical inactivity	32.5	1.82 (1.19–2.78)	21.0 % (5.8–36.6)	1,115 (308–1,942)
Smoking	20.6	1.59 (1.15–2.20)	10.8 % (3.0–19.8)	574 (159–1,050)
Low education	13.3	1.59 (1.35–1.86)	7.3 % (4.4–10.3)	386 (236–544)
Combined (maximum)			54.1 %	2 866 951[a]

[a] Absolute number
From Barnes and Yaffe (2011)

Obesity is associated with hypertension in one person but not in another; on average, "only" 50 % obese individuals have hypertension (Must et al. 1999). And yet both individuals develop Alzheimer's disease: in the obese-hypertensive person, dementia might be driven by vascular mechanisms, while in the obese-only individual, it might be due to axonal degeneration driven by the lack of androgens. Perhaps, the reason for this difference is the distribution of fat: more visceral fat in the hypertensive person and more subcutaneous fat in the normotensive individual. Such a difference in the relevant phenotype, together with a set of genetic (e.g. angiotensin or androgen receptor genes) and environmental (e.g. diet) risk factors, would allow us to determine which of the two pathogenic pathways need to be interrupted in order to delay the onset of Alzheimer's disease in a particular individual.

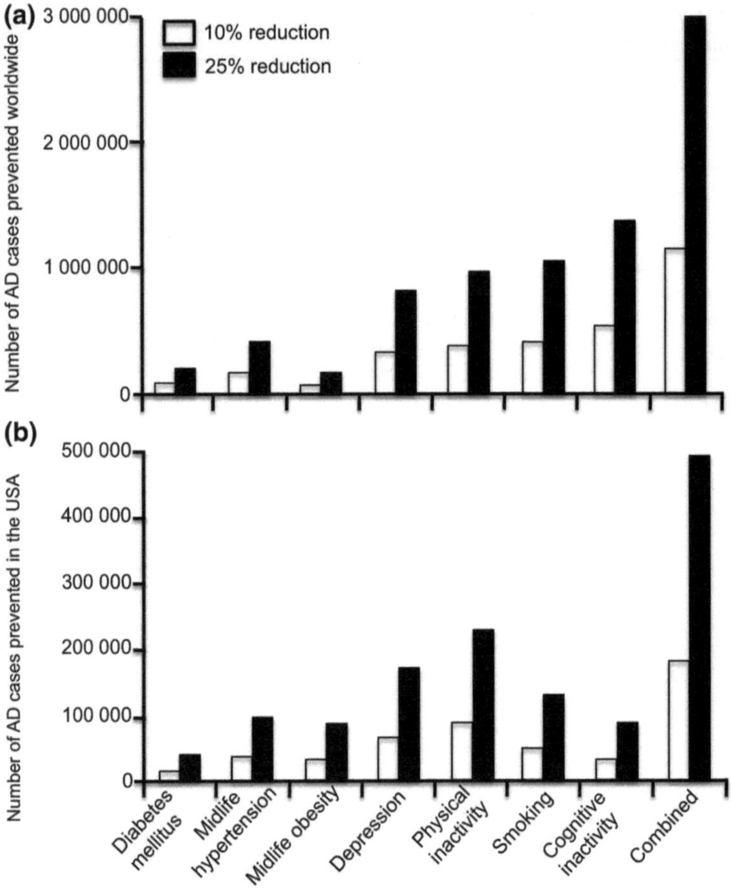

Fig. 10.8 Potential number of cases with Alzheimer's disease (AD) that could be prevented through risk factor reduction. The numbers of AD cases that could be potentially prevented (**a**) worldwide and (**b**) in the USA, through risk factor reductions of 10 or 25 %, was estimated by multiplying present prevalence estimates by 0.90 and 0.75, respectively, and subtracting the revised number of attributable cases from the original number. From Barnes and Yaffe (2011)

Then comes the intervention. This is when "knowledge is power" enters the equation: knowledge of the specific risk profile empowering the client and his/her physician to choose preventive strategy best suited for this profile. In our simplistic example, primary prevention suited for the obese-hypertensive person might involve attempts to reduce the amount of visceral fat (through diet or exercise). Secondary prevention would involve using medication to block signals generated by fat tissue that, in turn, increases blood pressure (Fig. 10.9), as well as manage the patient's hypertension though traditional anti-hypertensive medications. In the obese individual with mostly subcutaneous obesity, the above strategy is unlikely to work. Other than loosing weight, one would need to consider strategies leading

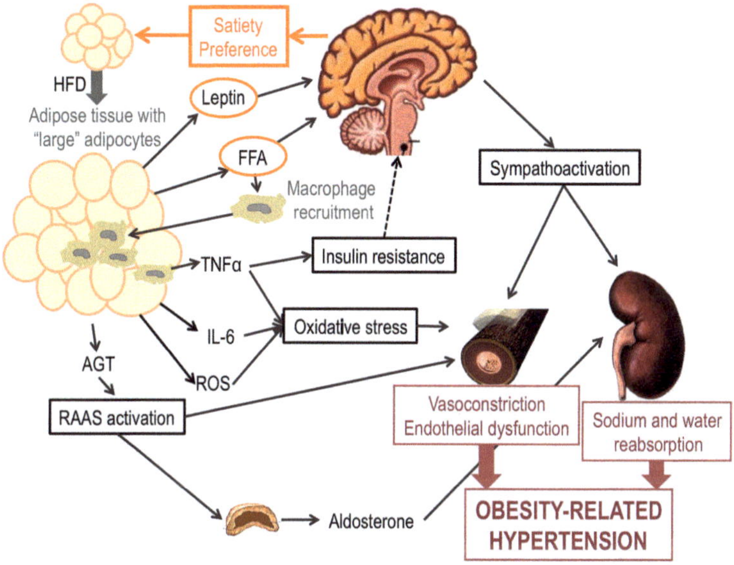

Fig. 10.9 Adipocyte: endocrine signals, autonomic activation and satiety. *HFD* high-fat diet; *FFA* free fatty acids; *AGT* angiotensinogen; *TNF-α* tumour necrosis factor-alpha; *IL-6* interleukin 6; *ROS* reactive oxygen species; *RAAS* rennin–angiotensin–aldosterone system; Based on Pausova (2006)

to an increase in androgen levels, whether through exercise (e.g. Okamoto et al. 2012) or medication.

It is important to note that these are only hypothetical examples: the multidimensional, multi-level approach to risk profiling described here is likely to generate much more nuanced strategies of personalized interventions, whether primary or secondary, and to discover additional mechanistic pathways leading to Alzheimer's disease in some individuals.

Why do we claim that "knowledge is power"? This statement refers not only to the informational value of the profile (as described above) but also, perhaps even more importantly, to its *motivational* value. Are people more likely to change their behaviour if provided with information about their personal risk? A few initial studies suggest that this is indeed the case (Roberts et al. 2005; Chao et al. 2008; Aspinwall et al. 2008). Thus, providing information about a specific genetic risk of Alzheimer disease (ApoE genotype) is 2.7 times more effective in changing health behaviours than providing risk estimates based on family history only (Text Box 10.5.).

Text Box 10.5. Knowing a gene variant changes health behaviours

In a randomized controlled study, 162 individuals with family history of Alzheimer's disease were randomized into two arms: (1) risk assessment based on the family history plus disclosure of ApoE genotype and (2) risk assessment based on the family history, without the ApoE genotype disclosure (Chao et al. 2008).A year later, individuals who learnt about having the risk allele were more likely to report improving their diet, taking vitamins/medications and exercising than individuals who knew that they do not have the risk allele or who did not receive any information about their ApoE genotype (Chao et al. 2008). The authors concluded that "that people who learn they are at high risk for AD are motivated to engage in behaviours to reduce risk, even if the effectiveness of activities is uncertain" (Chao et al. 2008, p. 97).

Overall, it is likely that new knowledge will provide the necessary foundation for personalized preventive medicine and, in the long run, reduce the burden associated with complex, chronic disorders of brain and body.

To conclude, population neuroscience—as introduced in this book—aspires to facilitate the dialogue between neuroscientists and epidemiologists, with geneticists providing the lingua franca of the trade. Traditional discipline-based science strives for depth and perfection. As a highly interdisciplinary enterprise, population neuroscience is broad in its scope, embracing a plurality of approaches in its quest for knowledge. I hope this book strikes the right balance between depth and breadth, thus making it easier for us to learn from each other.

References

Allison M (2012) Direct-to-consumer genomics reinvents itself. Nat Biotechnol 30(11): 1027–1029

Aspinwall LG, Leaf SL, Dola ER, Kohlmann W, Leachman SA (2008) CDKN2A/p16 genetic test reporting improves early detection intentions and practices in high-risk melanoma families. Cancer Epidemiol Biomarkers Prev 17(6):1510–1519. doi:10.1158/1055-9965.EPI-08-0010 17/6/1510 [pii]

Bajekal M (2005) Healthy life expectancy by area deprivation: magnitude and trends in England, 1994–1999. Health Stat Q 25:18–27

Barker DJ (2004) The developmental origins of well-being. Philos Trans R Soc Lond B Biol Sci 359(1449):1359–1366. doi:10.1098/rstb.2004.1518

Barnes DE, Yaffe K (2011) The projected effect of risk factor reduction on Alzheimer's disease prevalence. Lancet Neurol 10(9):819–828. doi:10.1016/S1474-4422(11)70072-2 S1474-4422(11)70072-2 [pii]

Biomarkers and surrogate endpoints: preferred definitions and conceptual framework (2001). Clin Pharmacol Ther 69(3):89–95. doi:10.1067/mcp.2001.113989

Broemeling AM, Watson DE, Prebtani F (2008) Population patterns of chronic health conditions, co-morbidity and healthcare use in Canada: implications for policy and practice. Healthc Q 11(3):70–76

Changing the Trajectory of Alzheimer's Disease: a national imperative. Alzheimer's Association (May 2010). http://www.alz.org/alzheimers_disease_trajectory.asp

Chao S, Roberts JS, Marteau TM, Silliman R, Cupples LA, Green RC (2008) Health behavior changes after genetic risk assessment for Alzheimer disease: the reveal study. Alzheimer Dis Assoc Disord 22(1):94–97. doi:10.1097/WAD.0b013e31815a9dcc

Chiu M, Austin PC, Manuel DG, Tu JV (2010) Comparison of cardiovascular risk profiles among ethnic groups using population health surveys between 1996 and 2007. CMAJ: Canadian Medical Association journal = journal de l'Association medicale canadienne 182(8):E301–E310. doi:10.1503/cmaj.091676

Hamburg MA, Collins FS (2010) The path to personalized medicine. N Engl J Med 363(4):301–304. doi:10.1056/NEJMp1006304

Hanson RL, Imperatore G, Bennett PH, Knowler WC (2002) Components of the "metabolic syndrome" and incidence of type 2 diabetes. Diabetes 51(10):3120–3127

Hastie T, Tishirani R, Friedman J (2001) The elements of statistical learning: data mining, inference, and prediction. Springer, New York

Janssens AC, van Duijn CM (2008) Genome-based prediction of common diseases: advances and prospects. Hum Mol Genet 17(R2):R166–R173. doi:10.1093/hmg/ddn250

Khaw KT, Wareham N, Bingham S, Welch A, Luben R, Day N (2008) Combined impact of health behaviours and mortality in men and women: the EPIC-Norfolk prospective population study. PLoS Med 5(1):e12. doi:10.1371/journal.pmed.0050012

Kohl HW 3rd, Craig CL, Lambert EV, Inoue S, Alkandari JR, Leetongin G, Kahlmeier S (2012) The pandemic of physical inactivity: global action for public health. Lancet 380(9838):294–305. doi:10.1016/S0140-6736(12)60898-8 S0140-6736(12)60898-8 [pii]

Krishnan KR (2005) Psychiatric and medical comorbidities of bipolar disorder. Psychosom Med 67(1):1–8. doi:10.1097/01.psy.0000151489.36347.18 67/1/1 [pii]

Limdi NA, Arnett DK, Goldstein JA, Beasley TM, McGwin G, Adler BK, Acton RT (2008) Influence of CYP2C9 and VKORC1 on warfarin dose, anticoagulation attainment and maintenance among European-Americans and African-Americans. Pharmacogenomics 9(5):511–526. doi:10.2217/14622416.9.5.511

Lopez AD, Mathers CD, Ezzati M, Jamison DT, Murray CJ (2006) Global and regional burden of disease and risk factors, 2001: systematic analysis of population health data. Lancet 367(9524):1747–1757. doi:10.1016/S0140-6736(06)68770-9

Lusis AJ, Attie AD, Reue K (2008) Metabolic syndrome: from epidemiology to systems biology. Nat Rev Genet 9(11):819–830. doi:10.1038/nrg2468

Mathers CD, Iburg KM, Salomon JA, Tandon A, Chatterji S, Ustun B et al (2004) Global patterns of healthy life expectancy in the year 2002. BMC Public Health 4:66. doi:10.1186/1471-2458-4-66

Maud PJ, Pollock ML, Foster C, Anholm JD, Guten G, Al-Nouri M, Hellman C, Schmidt DH (1981) Fifty years of training and competition in the marathon: Wally Hayward, age 70–a physiological profile. S Afr Med J 59(5):153–157

McInnes GT, Semple PF (1994) Hypertension: investigation, assessment and diagnosis. Br Med Bull 50(2):443–459

Meda SA, Narayanan B, Liu J, Perrone-Bizzozero NI, Stevens MC, Calhoun VD, Glahn DC, Shen L, Risacher SL, Saykin AJ, Pearlson GD (2012) A large scale multivariate parallel ICA method reveals novel imaging-genetic relationships for Alzheimer's disease in the ADNI cohort. Neuroimage 60(3):1608–1621. doi:10.1016/j.neuroimage.2011.12.076 S1053-8119(12)00017-1 [pii]

Moore SC, Patel AV, Matthews CE, Berrington de Gonzalez A, Park Y, Katki HA, Linet MS, Weiderpass E, Visvanathan K, Helzlsouer KJ, Thun M, Gapstur SM, Hartge P, Lee IM (2012) Leisure time physical activity of moderate to vigorous intensity and mortality: a large pooled cohort analysis. PLoS Med 9(11):e1001335. doi:10.1371/journal.pmed.1001335 PMEDICINE-D-12-00949 [pii]

Must A, Spadano J, Coakley EH, Field AE, Colditz G, Dietz WH (1999) The disease burden associated with overweight and obesity. JAMA 282(16):1523–1529

Okamoto M, Hojo Y, Inoue K, Matsui T, Kawato S, McEwen BS, Soya H (2012) Mild exercise increases dihydrotestosterone in hippocampus providing evidence for androgenic mediation of neurogenesis. Proc Natl Acad Sci U S A 109(32):13100–13105. doi:10.1073/pnas.1210023109 1210023109 [pii]
Osby U, Brandt L, Correia N, Ekbom A, Sparen P (2001) Excess mortality in bipolar and unipolar disorder in Sweden. Arch Gen Psychiatry 58(9):844–850
Pausova Z (2006) From big fat cells to high blood pressure: a pathway to obesity-associated hypertension. Curr Opin Nephrol Hypertens 15(2):173–178. doi:10.1097/01.mnh.0000214775.42103.a5
Program NCPE (2010) Risk Assessment Tool for Estimating 10-year Risk of Developing Hard CHD (Myocardial Infarction and Coronary Death). http://hp2010.nhlbihin.net/atpiii/calculator.asp?usertype=prof
Pyne JM, Fortney JC, Tripathi S, Feeny D, Ubel P, Brazier J (2009) How bad is depression? Preference score estimates from depressed patients and the general population. Health Serv Res 44(4):1406–1423. doi:10.1111/j.1475-6773.2009.00974.x
Roberts JS, Cupples LA, Relkin NR, Whitehouse PJ, Green RC (2005) Genetic risk assessment for adult children of people with Alzheimer's disease: the risk evaluation and education for Alzheimer's disease (REVEAL) study. J Geriatr Psychiatry Neurol 18(4):250–255. doi:10.1177/0891988705281883 18/4/250 [pii]
Rose G (2008) Rose's strategy of preventive medicine. Oxford University Press, Oxford
Scherer HU, Dorner T, Burmester GR (2010) Patient-tailored therapy in rheumatoid arthritis: an editorial review. Curr Opin Rheumatol 22(3):237–245. doi:10.1097/BOR.0b013e328337b832
Schuckit MA (2006) Comorbidity between substance use disorders and psychiatric conditions. Addiction 101(Suppl 1):76–88. doi:10.1111/j.1360-0443.2006.01592.x ADD1592 [pii]
Third Report of the National Cholesterol Education Program (NCEP) expert panel on detection, evaluation, and treatment of high blood cholesterol in adults (adult treatment panel III) final report (2002). Circulation 106(25):3143–3421
Thompson JB, Rivera JJ, Blumenthal RS, Danyi P (2006) Primary prevention for patients with intermediate Framingham risk scores. Curr Cardiol Rep 8(4):261–266
Tremblay MS, LeBlanc AG, Kho ME, Saunders TJ, Larouche R, Colley RC, Goldfield G, Gorber SC (2011) Systematic review of sedentary behaviour and health indicators in school-aged children and youth. Int J Behav Nutr Phys Act 8:98. doi:10.1186/1479-5868-8-98 1479-5868-8-98 [pii]
Tsuno N, Homma A (2009) What is the association between depression and Alzheimer's disease? Expert Rev Neurother 9(11):1667–1676. doi:10.1586/ern.09.106
Vanhanen M, Koivisto K, Moilanen L, Helkala EL, Hanninen T, Soininen H, Kervinen K, Kesaniemi YA, Laakso M, Kuusisto J (2006) Association of metabolic syndrome with Alzheimer disease: a population-based study. Neurology 67(5):843–847. doi:10.1212/01.wnl.0000234037.91185.99
Vineis P, Schulte P, McMichael AJ (2001) Misconceptions about the use of genetic tests in populations. Lancet 357(9257):709–712. doi:10.1016/S0140-6736(00)04136-2

Anonymous

Biomarkers and surrogate endpoints: preferred definitions and conceptual framework (2001)
Third Report of the National Cholesterol Education Program (NCEP) expert panel on detection, evaluation, and treatment of high blood cholesterol in adults (adult treatment panel III) final report (2002)
Program NCPE (2010). Risk Assessment Tool for Estimating 10-year Risk of Developing Hard CHD (Myocardial Infarction and Coronary Death)
Changing the Trajectory of Alzheimer's Disease: a national imperative. Alzheimer's association (May 2010)

Index

T. Paus, *Population Neuroscience*, DOI: 10.1007/978-3-642-36450-1,

N

MIX
Papier aus verantwortungsvollen Quellen
Paper from responsible sources
FSC® C105338

If you have any concerns about our products,
you can contact us on
ProductSafety@springernature.com

In case Publisher is established outside the EU,
the EU authorized representative is:
Springer Nature Customer Service Center GmbH
Europaplatz 3, 69115 Heidelberg, Germany

Printed by Libri Plureos GmbH
in Hamburg, Germany